Environmental Water

Environmental Water

Advances in Treatment, Remediation and Recycling

V.K. Gupta FNASc, FWIF
Professor and Head
Department of Chemistry
Indian Institute of Technology,
Roorkee, India

I. Ali
Department of Chemistry
Jamia Millia Islamia (Central University),
India

ELSEVIER

Amsterdam • Boston • Heidelberg • London • New York • Oxford
Paris • San Diego • San Francisco • Singapore • Sydney • Tokyo

Elsevier
Radarweg 29, PO Box 211, 1000 AE Amsterdam, The Netherlands
The Boulevard, Langford Lane, Kidlington, Oxford OX5 1GB, UK

First edition 2013

British Library Cataloguing in Publication Data
A catalogue record for this book is available from the British Library

Library of Congress Cataloging-in-Publication Data
A catalog record for this book is available from the Library of Congress

Printed in the United States of America
Transferred to Digital Printing, 2012

ISBN: 978-0-444-59399-3

Dedications

Dedicated to the memories of my late father
Shri Jeevan Lal Gupta s/o late Shri Sita Ram ji
V.&PO Lakhnauti Distt. Saharanpur(UP)

Vinod Kumar Gupta

Dedicated to the memories of my late parents:
Basheer Ahmed and Mehmudan Begum.

Imran Ali

Contents

Preface

The global population is increasing; it is estimated that the world population will increase to 7.9 billion by the end of 2020. Rapid industrialization, modern methods of agricultural, domestic activities and other geological, environmental and global changes have put undue pressure on the demand and quality of fresh water. According to United Nations World Water Development Report, some 2 million tons of waste are discharged to the water bodies per day including industrial wastes, dyes and chemicals, human waste and agricultural wastes (fertilizers and pesticides). Because of this, the world may be under great fresh water scarcity. Fresh water is an important and essential component of this universe and plays a vital role in the proper functioning of the Earth's ecosystem. Our water resources are limited and, hence, water treatment and recycling methods are the only alternatives for getting fresh water in the coming decades. Therefore, there is a great need for the development of a suitable, inexpensive and rapid wastewater treatment technique and reuse or conservation methods in the present century. Several water treatment technologies have been developed for this purpose. In view of these facts, this book deals with a detailed discussion on the different types of water treatment and recycling techniques in terms of their basic principles, applications, maintenance and suitability. The book has been divided into eight chapters; each chapter has been exclusively dedicated to a particular treatment technique, highlighting the principles, mechanism, operating parameters, ways of maximizing efficiency, advantages, disadvantages and future perspectives. The technologies discussed include adsorption, reverse osmosis, membrane filtration, electrical, biological, remediation and recycling. A systematic approach of water treatment and recycling involving their understanding, evaluation and selection parameters has been presented. A comparison of the technologies has been presented by discussing their performance, sludge production, life period and operations. Moreover, optimization strategies for each treatment technique is explained which will be helpful to design future experiments in this area. This book is important and unique because it is one of the first texts that deals with the global discussions on water scarcity solutions. The purpose of this book is to provide the guidelines for the selection of the technologies or their combinations for various applications so that one can select the exact and correct technology. The book will have immense utility to scientists, researchers, academicians and students working in this area. A brief guideline for the selection of the appropriate technologies for specific applications has been evaluated which will be beneficial to the regulatory authorities for formulating the regulations and legislations of water treatment.

Vinod Kumar Gupta
Imran Ali

Acknowledgments

I am thankful to my institution and department for permitting me to write this book. It was difficult task without the cooperation of my mother, Smt. Kiran Devi, wife Prerna, son and daughter-in-law; Rajat and Shilpi; daughter and son-in-law, Vartika and Abhinav and my dearest granddaughters, Shiriya and Sharanya.

I thank Ms Arunima, my research student, who has helped me from time to time to accomplish this task.

V. K. Gupta

It was indeed a difficult task for me to complete this book but the extreme help and cooperation of my wife Seema Imran made it reality. My lovely and sweet thank is also to my dearest son Al-Arsh Basheer Baichain, who has given me freshness and fragrance continuously during the completion of this difficult task. I would also like to acknowledge my other family members, relatives who have helped me directly and indirectly during this period.

I also pay my sincere thanks to Prof. Kishwar Saleem and Prof. Tabrez A. Khan, Chemistry Department, Jamia Millia Islamia, New Delhi for their moral support.

I. Ali

About the Authors

Dr. Vinod Kumar Gupta, **Professor and Head, Department of Chemistry** has an outstanding academic record. He was a Visiting Professor (DAAD) at the Technical University of Chemnitz and Freie University of Berlin and KFUPM Chair Professor at King Fahd University of Petroleum and Minerals, Dhahran, Saudi Arabia. He has visited more than **20 countries** for collaborative research and academic purposes.

Prof. Gupta has more than **33 years** experience of teaching and research and is a sought after speaker in the area of environmental chemistry and chemical sensors. He has delivered invited keynote lectures in many conferences and institutes in India and abroad and has guided **33 Ph.D** theses, **10 M.Phil.** and **33 M.Sc. dissertations.** He has published **one book on—Nano Chromatography and Capillary Electrophoresis: Pharmaceutical and Environmental Analyses, John Wiley & Sons, Inc. (2009)** and more than **325 research papers** in international journals of very high impact factors and presented more than **50 papers** in conferences in India and abroad. He has contributed **12 monographs** in various books and encyclopedias and has published **23 reviews** on topics of contemporary significance. His publications have received more than **12,000** citations with an h **index = 70**. He is one among the **top most highly cited scientists and engineers of India** (isihighlycited.com). He has undertaken many sponsored research and consultancy projects in India and abroad. **He is an Editor of Environmental Science and Pollution Research (published by Springer; IF = 2.87)** and is on the Editorial Board of many international and national journals.

Prof. Gupta has received a number of awards and honors to name few; **Marie Curie** Fellowship of European Commission, 1992; Khosla research Prize three times; India Citation Laureate Award, 2004 by Thomson Scientific, USA; Fellow of the World Innovation Foundation (FWIF) 2004; Fellow of the National Academy of Sciences (FNASc) 2008; Cited among the 25 top most scientists from India as per India Today (19 September 2011).

Dr. Gupta has wide experience of administration also and is known for his disciplined and transparent approach toward administration. He has occupied almost all important administrative positions in the Department and Institute, namely; **Associate Dean of Students' Welfare, where he looked after the management of all the hostels of IIT Roorkee (January 2003 to December 2005); Dean of Students' Welfare (1 January 2006 to December 2008). Tenure as DOSW has been highly successful and incident free; Member of Syndicate and Senate of the erstwhile University of Roorkee; Member of the first BoG of IIT Roorkee; Head of the Chemistry Department wef 01-01-2011.**

Dr. Imran Ali is currently an academician and researcher at Department of Chemistry, Jamia Millia Islamia (Central University), Jamia Nagar, New Delhi—110025, India. He obtained his M.Sc. (1986) and Ph.D. (1990) degrees from Indian Institute of Technology, Roorkee, India. His research areas are water treatment and analyses of chiral and simple molecules in biological and the environmental matrices.

Dr. Ali is author and coauthor of more than 250 articles including research papers, reviews, book and encyclopedia chapters, technical reports and conference presentations. He has papers in Nature and Chemical Reviews (ACS), and is an author of four books viz. "*Chiral Separations by Liquid Chromatography and Related Technologies*" published by Marcel Dekker, Inc., New York, USA (2003); "*Chiral pollutants: Distribution, toxicity and analysis by chromatography and capillary electrophoresis*" published by John Wily & Sons, Chichester, UK (2004); "*Instrumental methods in metal ions speciation: Chromatography, Capillary Electrophoresis and Electrochemistry*" published by Taylor & Francis Ltd., New York, USA (2006) and "*Nano chromatography and capillary electrophoresis: pharmaceutical and environmental analyses*", published by Wiley & Sons, Hoboken, USA. He is an editor of ScienceJet journal and member of the editorial boards of various journals. He is a life member of various national and international scientific societies such as Chromedia News, Chromedia Chromatographers Community, Amsterdam, The Netherlands, Indian Science Congress Association, Indian Society of Analytical Scientists, Chromatographic Society of India, Indian Association for Cancer Research and American Nano Society. Dr Ali is also a research advisor and Deputy Director General of American Biographical Institute, Inc., USA. Dr Imran Ali has research collaboration in more than 15 countries all over the world and also a widely traveled worker.

1 Introduction

CHAPTER OUTLINE

1.1 The Environment

"A sound mind exists in a sound body" is a well-known proverb in English literature and the same is true for our environment. A clean environment is essential for the proper growth, health, and the persistence of the human beings and other organisms. The conservation and the protection of the environment are essential in the present industrialized and developing world. Over growth of the population in some countries is one of the most pressing problems of our age. The pollution problem of the environment has now reached at a level that possesses a potential threat not only to the health but also to the entire populations. The quality of our environment is deteriorating continuously due to the accumulation of undesirable constituents into it. The main sources of the contamination are industrialization, domestic activities, agricultural activities, atomic explosions, and other environmental and global changes. These activities and changes, if

Environmental Water, http://dx.doi.org/10.1016/B978-0-444-59399-3.00001-5

improperly controlled, can destroy the quality of our environment. Broadly, the environment is divided into three parts, i.e. atmosphere including the air sphere around the earth, lithosphere which consists of the earth itself, and the hydrosphere, i.e. all the water bodies including ocean, surface, and groundwater. Hydrosphere and the atmosphere components of the environment are directly and readily available for the contamination by the pollutants. Therefore, the quality of these components of the environment is deteriorating continuously which is a matter of serious thinking and worried. Again, the notorious pollutants find their ways easily through water bodies, while atmosphere is being contaminating only by some gases and volatile organic pollutants. Therefore, the ground and surface water at some places in the world are not suitable for drinking purpose due to the presence of aesthetic and toxic pollutants. Similarly, the quality of atmosphere at some cities and industrial areas of the world is not safe for health. The importance of the conservation and improvement of the environment is essential and urgent need [1–3]. Environmental awareness has grown dramatically in past decades and several nations are taking lead in implementing the laws related to the environment. In view of this, the environmental authorities are asking the data and information on the pollution level and the improvement measures to control the contamination of the environment.

1.2 Water Resources

Water resources are defined as the sources which produce the water for our different types of uses and also those sources that give the huge benefit to the life of the humans. The water which is used in the production of different types of useful products are also included in the water resources. Basically, the function of the water resources is to overcome the desires or the requirement of the water for the agricultural or household purposes.

Water resources are one of the most essential issues due to climate change impacts and adaptation. A clean and reliable water supply is critical for domestic use, food and energy production, transportation, recreation, and maintenance of natural ecosystems [4]. Different types of water resources are used for the production of water around the world and they are divided into different categories (Figure 1-1.) based on their composition and also on the basis of their uses for the benefit of the humanity. Some important types of water resources are as follows.

1.2.1 Saltwater Resources

Water contains dissolved chemical materials called “salts.” When the concentration of these dissolved materials becomes great, the water is referred to as saltwater or saline water.

Saltwater resources are the types of water resources that can make the saltwater for the sake of the benefit of the mankind and also save the environment from harm. Saltwater

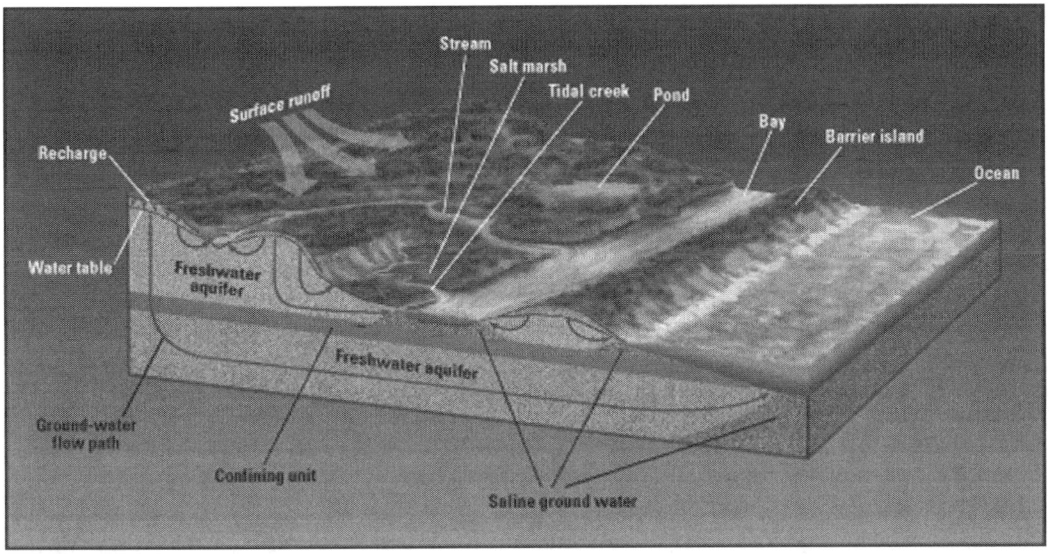

FIGURE 1-1 Water resources. For color version of this figure, the reader is referred to the online version of this book.

resources are also useful for generating the hydroelectric energy. Desalination of water separates the water into freshwater and saline water. Main saltwater resources are ocean, seawater, etc.

a. Ocean: Oceans are the largest single source of salts in the atmosphere, and sodium and chloride are the most abundant ions in air masses over the sea [5–7]. Chloride and sodium concentrations, therefore, are high in air masses near seacoasts but decrease rapidly with increasing distance inland. These airborne salts are delivered to coastal watersheds by precipitation. Chloride concentrations in precipitation, however, are relatively small compared to seawater. Average chloride concentrations in precipitation measured at about two-dozen atmospheric deposition monitoring stations along the Atlantic coast in 2000, for example, ranged from less than 0.2 to 2.8 mg/L [7]. Concentrations of sodium and chloride can be increased in soils, shallow surface waters (such as tidal lagoons), and groundwater by evaporation and evapotranspiration.

b. Sea water: Tides and storm surges, which can be local sources of increased groundwater salinity in low-lying coastal areas and fossil seawater in unflushed parts of an euifer: Such water either was trapped in sedimentary formations when they were deposited (connate water) or flowed into the formations during periods of relatively high sea levels when seawater flooded low-lying coastal areas [8–12].

c. Some of the ocean's salts have been dissolved from rocks and sediments below the sea floor [13,14].

d. Others have escaped from the Earth's crust through volcanic vents as solid and gaseous materials [15].

1.2.2 Surface Water Resources

Another type of water resources is the surface water resource in which the water is present in the rivers or in the streams. Basically, this type of water is used in many useful purposes such as for the industrial use, for agricultural use, and for the generation of different types of energy, i.e. hydroelectrical energy. Surface water is very important because its 90–95% is used in the industry for manufacturing of different products [16,17]. Surface water mainly occurring in lakes, rivers, streams, or other freshwater sources used for drinking water supplies. While most drinking water in the United States is withdrawn from groundwater sources, surface water remains a significant water resource [18].

Surface water is naturally replenished by precipitation and naturally lost through discharge to evaporation and subsurface seepage into the ground. Although there are other sources of groundwater, such as connate water and magmatic water, precipitation is the major one and groundwater originated in this way is called meteoric water [19].

1.2.3 Groundwater Resources

Groundwater is generally stored in aqueducts, underground layers of porous rocks that are saturated with water. These aqueducts receive water as soil becomes saturated with precipitation or through stream and river runoff. As the aqueducts exceed their capacity for water storage, they will bleed water back into streams or rivers. The aqueducts maintain a natural balance of water, alternately receiving or giving water as their saturation levels oscillate. Throughout this process, water constantly moves between surface and groundwater sources, sharing contaminants as shown in Figure 1-2. Groundwater divided into two zones, i.e. saturated and unsaturated and these zones are separated with the help of water table below the ground [20,21].

1.3 Environmental Pollutants

Pollution refers to the entry of contaminants into natural environment that causes disorder, instability, and harm or discomfort to the ecosystem [22]. Pollutants, the causative factors of pollution, can be either foreign substances or naturally occurring contaminants. Environmental pollution is a major issue of concern and it is likely to influence the health of human populations [23]. In addition to the effects of environment pollution on human beings, it also affects plants, trees, and animals. Environment pollution is a global problem and its potential to influence the health of human populations is great [24]. The significance of environmental factors to the health and well-being of human populations is increasingly apparent [25]. The most serious hazardous effects of environmental pollution can be seen in the densely settled urban–industrial centers of the more developed countries [26]. More than 80% polluted water have been used for irrigation purposes [27]. Industry, clustered in urban and semi-urban areas surrounded by densely populated, low-income localities, continues to pollute the environment [28]. The present scenario is in favor of the fact that human exposure to

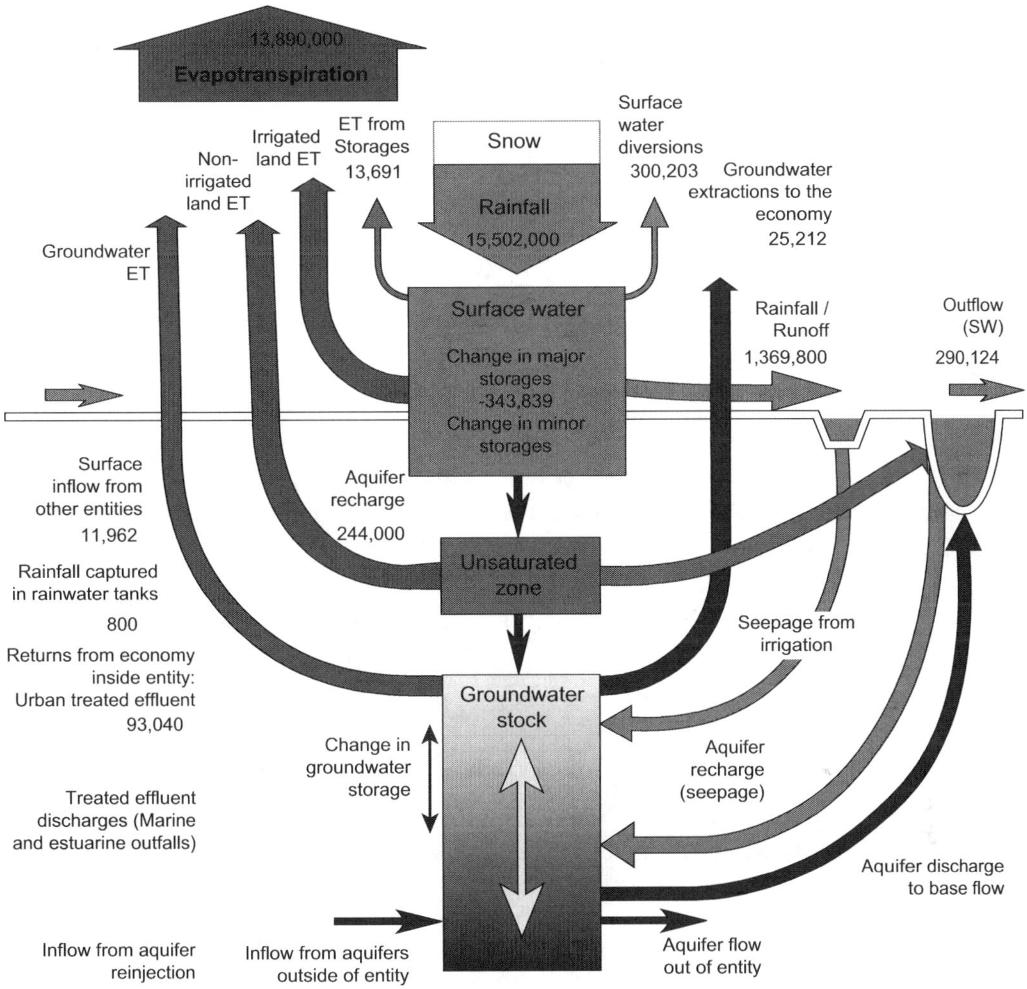

FIGURE 1-2 Water cycle. For color version of this figure, the reader is referred to the online version of this book.

pollution is believed to be more intense now than at any other time in human existence [29]. Pollution can be made by human activity as well as by natural forces [24]. Moreover, selfish private enterprise and their lack of awareness of public well-being and social costs [30] and natural disasters [31], e.g. volcanic ash from Iceland [32] are the other main contributing factors of pollution.

Water pollution is a major global problem which requires ongoing evaluation and revision of water resource policy at all levels. It has been suggested that it is the leading worldwide cause of diseases and deaths [33]. Water pollution accounts for the deaths of more than 14,000 people daily [34]. In addition to the acute problems of water pollution in developing countries, developed countries continue to struggle with pollution problems as well. In the most recent national report on water quality in the United States, 45% of assessed stream miles, 47% of assessed lake acres, and 32% of assessed bay and estuarine square miles have been classified as polluted [35].

Water is mainly referred to as polluted when it is impaired by anthropogenic contaminants and becomes unable to support a human use, such as drinking water, and/or undergoes a marked shift in its ability to support its constituent biotic communities, such as fish. However, natural phenomena like volcanoes, algae blooms, storms, and earthquakes also cause major changes in water quality and the ecological status of water.

1.4 Types of Pollutants

Pollutants have been categorized into organic, inorganic, and biological on the basis of their nature and origin. A classification of pollutants is shown in Figure 1-3. Besides, a list of industries discharging various contaminants is given in Table 1-1. A brief summary of the types of pollutants is discussed below:

1.4.1 Organic Pollutants

Agents that degrade water quality and are organic in nature form big threats to healthy human existence and aquatic life. Organic water pollutants mainly include

1. Food processing waste, which can include oxygen-demanding substances, fats, and grease.
2. Petroleum hydrocarbons, including fuels (gasoline, diesel fuel, jet fuels, and fuel oil), lubricants (motor oil), and fuel combustion by-products [36].
3. Disinfection by-products (DBPs) found in chemically disinfected drinking water, such as chloroform.
4. Volatile organic compounds (VOCs), such as industrial solvents, from improper storage.
5. Insecticides and herbicides, a huge range of organohalides and other chemical compounds.
6. Tree and bush debris from logging operations.
7. Chlorinated solvents, which are dense nonaqueous phase liquids, may fall to the bottom of reservoirs, since they do not mix well with water and are denser.
8. Perchlorate.
9. Polychlorinated biphenyl (PCB).
10. Trichloroethylene.

Persistent organic pollutants (POPs) are a class of organic pollutants, resistant to environmental degradation. These are quite stable organic compounds and therefore, easily resist degradation to different environmental stresses. Because of this, they have been observed to persist in the environment, to be capable of long-range transport,

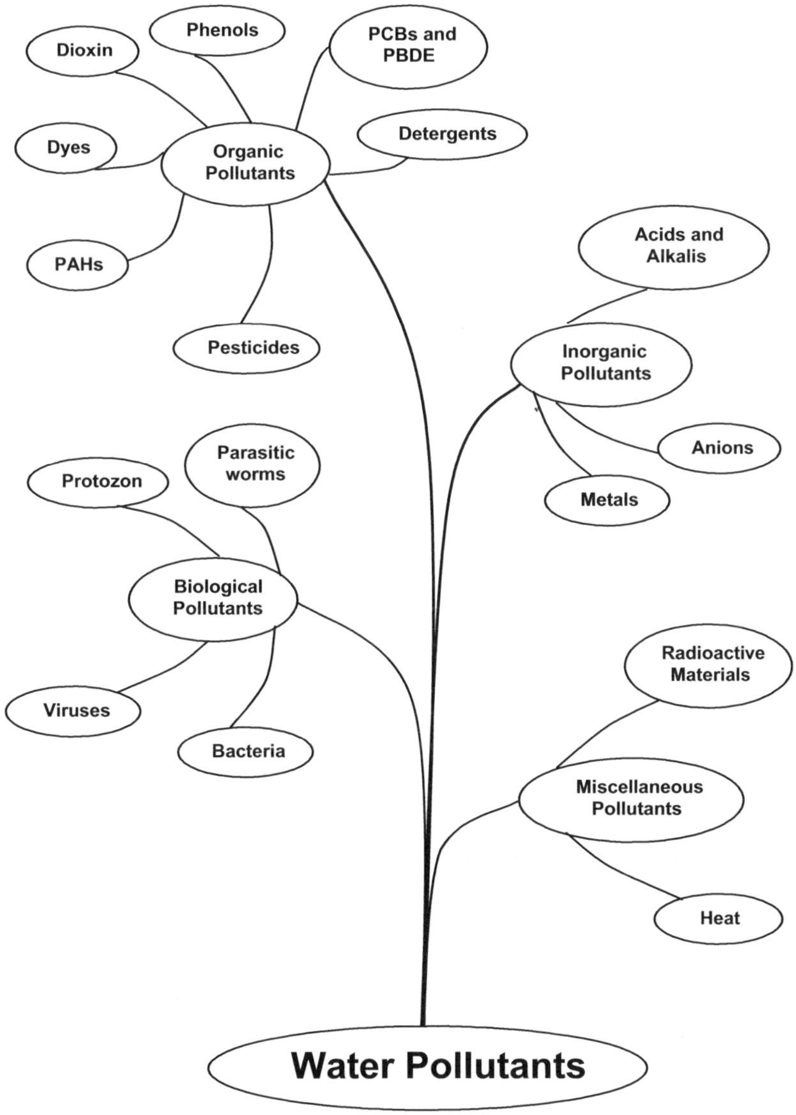

FIGURE 1-3 A classification of water pollutants on the basis of their chemical nature.

bioaccumulation in human and animal tissue and biomagnifying in food chains and to have potential significant impacts on human health and the environment. Many POPs are currently used as pesticides. Others are used in industrial processes and in the production of a range of goods such as solvents, polyvinyl chloride, and pharmaceuticals. There are a few natural sources of POPs but most POPs are created by humans in industrial processes, either intentionally or as by-products [37]. The United Nations Environment

Table 1-1 List of Industries Discharging Various Contaminants

	Industries	Inorganic Pollutants	Organic Pollutants
1	Refineries	Sulphides, Hg, Cr, Sb	Oil-30, phenol-30, dioxin, furan, PAH, PCBs
2	Production and processing of metals	As, Cd, Pb, Se, Fe, Zn, Al, Ni	PCB, PAH, dioxins, oil, cyanides
3	Glass manufacture	Scrap silicates and other mineral oxides	Emulsified oil, soda ash
4	Ceramics industry	Fluorides	Not available
5	Coke–chemical industries	Ammonia, cyanides	Phenols, hydrogen sulphide, tars, hydrocarbons, thiocyanates, organic suspension
6	Incineration and landfills	Cadmium, lead, mercury	Chlorobenzene, chlorophenols, furans, and dioxins
7	Pulp & Paper, Board Mills	Sulphites, free chlorine	Methyl mercaptans, pentachlorophenol aldehydes, ketones, lignite sulphonate, and organic suspension
8	Organic chemicals	Not available	Phenols, organic acids, nitro-compounds
9	Fertilizers	Sulfuric acid, phospsoric, and fluosilicic acids, fluorine compounds, HCl, As	Monoethanol amines
10	Petroleum refining	Sulphides, chlorides, sulfates	Naphthalene acid, petroleum products, phenols, surfactants
11	Steel mills	NH_3/N-1000	Oil, grease, phenols 1300
12	Metal industry	Copper, lead, cadmium, chromium, zinc	Not available
13	Textile industry	Cr-5	Dyes, detergents
14	Iron and steel industry	Cyanogen	Coke, limestone, phenol, oils
15	Leather tanning and finishing	Chromium in the form of chromate, ammonical nitrogen, sodium chloride	Precipitated lime
16	Polymer and plastic industry	Sulfates	Styrene, acetonitrile, acrylates, phenols, aromatic hydrocarbons, aldehydes, cyclohexane, organic acids, and suspension
17	Synthetic rubber industry	Not available	Butylenes, butadiene, organic acids, acetonitrile, aldehydes, alcohols, and hydrocarbons
18	Chlorine industry	Mercury, chlorine, and chlorides	Not available
19	Distillery waste industry	Sodium, potassium, calcium, iron, sulfates, and chlorides	Not available
20	Soap and detergent industry	Not available	Propylene-derived alkyl benzene sulphonate
21	Photographic industry	Hypo, silver bromide spent solution, and fixing reducing agents	Not available

Table 1-1 List of Industries Discharging Various Contaminants—Cont'd

	Industries	Inorganic Pollutants	Organic Pollutants
22	Food processing industry	Alkali metals, phosphates	Amino acids
23	Pesticide industry		Dichlorophenol
24	Tannery industry	Chromium, sulphides	Phenols, tartaric acid
25	Explosive industry	Data not available	Nitro-compounds
26	Laundry industry	Alkalies, free chlorine	Oils, grease
27	Wood processing industry	Zinc, sulphides, and alkalies	Not available

Programme Governing Council in 1995, decided to begin investigating POPs. They initially began with a short list of the following 12 POPs, whom they referred as the "dirty dozen" [38]: some of which are as follows (a) aldrin, (b) benzenehexachloride, (c) mirex, (d) toxaphene, (e) chlordane, (f) DDT, (g) dieldrin, (h) endrin, and (i) heptachlor. The structures of some organic pollutants are given in Figure 1-4.

Since then, this list has generally been accepted to include such substances as carcinogenic polycyclic aromatic hydrocarbons (PAHs) and certain brominated flame retardants, as well as some organometallic compounds such as tributyltin (TBT). The groups of compounds that make up POPs are also classed as PBTs (Persistent, Bio-accumulative, and Toxic) or TOMPs (Toxic Organic Micro Pollutants). Some of the chemical characteristics of POPs include low water solubility, high lipid solubility, semi-volatility [37], and high molecular masses. POPs with molecular masses lower than 236 g/mol are less toxic, less persistent in the environment, and have more reversible effects than those with higher molecular masses [37]. POPs are frequently halogenated, usually with chlorine [37]. The more chlorine groups a POP has, the more resistant it is to being broken down over time [37]. One important factor of their chemical properties such as lipid solubility results in the ability to pass through biological phospholipids membranes and bioaccumulate in the fatty tissues of living organisms [37].

POP exposure can cause death and illnesses including disruption of the endocrine, reproductive, and immune systems, neurobehavioral disorders, and cancers possibly including breast cancer. Exposure to POPs can take place through diet, environmental exposure, or accidents [37]. A study published in 2006 indicated a link between blood serum levels of POPs and diabetes [39]. Individuals with elevated levels of POPs (DDT, dioxins, PCBs, and chlordane, among others) in their body were found to be up to 38 times more likely to be insulin resistant than individuals with low levels of these pollutants, though the study did not demonstrate a cause and effect relationship [39]. As most exposure to POPs is through consumption of animal fats, study participants with high levels of serum POPs are also very likely to be consumers of high amounts of animal fats, and thus, the consumption of the fats themselves or other associated factors may be responsible for the observed increase in insulin resistance. Another possibility is that insulin resistance causes increased accumulation of POPs.

1.4.2 Inorganic Pollutants

Inorganic chemical pollutants are naturally found in the environment but due to human development, these pollutants are often concentrated and released into the environment in urban storm water. The primary inorganic pollutants of concern in urban storm water are cadmium, copper, lead, zinc, nitrogen, nitrate, nitrite, ammonia, phosphorous, and phosphate. These chemicals are used in every aspect of human activity and are often highly toxic to humans and the environment. Some common trace metal ions such as cadmium (Cd), copper (Cu), lead (Pb), zinc (Zn), nitrogen (N), phosphorous (P), nitrate, nitrite, ammonia, and phosphate are present in the environment as a inorganic pollutants. Some of the events that lead to the increased accumulation of inorganic pollutants into the environment include

1. Acidity caused by industrial discharges (especially sulfur dioxide from power plants).
2. Fertilizers containing nutrients—nitrates and phosphates—which are found in storm water runoff from agriculture, as well as commercial and residential use.
3. Chemical waste as industrial by-products.
4. Ammonia from food processing waste.
5. Silt (sediment) in runoff from construction sites, logging, slash. and burn practices or land clearing sites.
6. Heavy metals from motor vehicles (via urban storm water runoff) and acid mine drainage.

1.4.3 Biological Pollutants

Biological contaminants include bacteria, molds, mildew, cat saliva, viruses, animal dander, house dust, mites, cockroaches, and pollen. Such pollutants originate from different sources. Pollens originate from plants; viruses are transmitted by people and animals; bacteria are carried by people, animals, and soil and plant debris; and household pets are sources of saliva and animal dander. The protein in urine from rats and mice is a potent allergen. When it dries, it can become airborne. Contaminated central air handling systems can become breeding grounds for mold, mildew, and other sources of biological contaminants and can then distribute these contaminants through the home [40]. Molds, mildew, fungi, bacteria, and dust mites are some of the main biological pollutants inside the house. Some, such as pollen, are generated outside the home. Mold and mildew are generated in the home and release spores into the air. Mold, mildew, fungi, and bacteria are often found in areas of the home that have high humidity levels, such as bathrooms, kitchens, laundry rooms, or basements. Dust mites and animal dander are problematic when they become airborne during vacuuming, making beds, or when textiles are disturbed [41,42]. Outdoor air pollution in cities is a major health problem. Much efforts and money continue to be spent cleaning up pollution in the outdoor air. Many ordinary activities such as cooking, heating, cooling, cleaning, and redecorating can cause the release and spread of indoor pollutants at home and thereby,

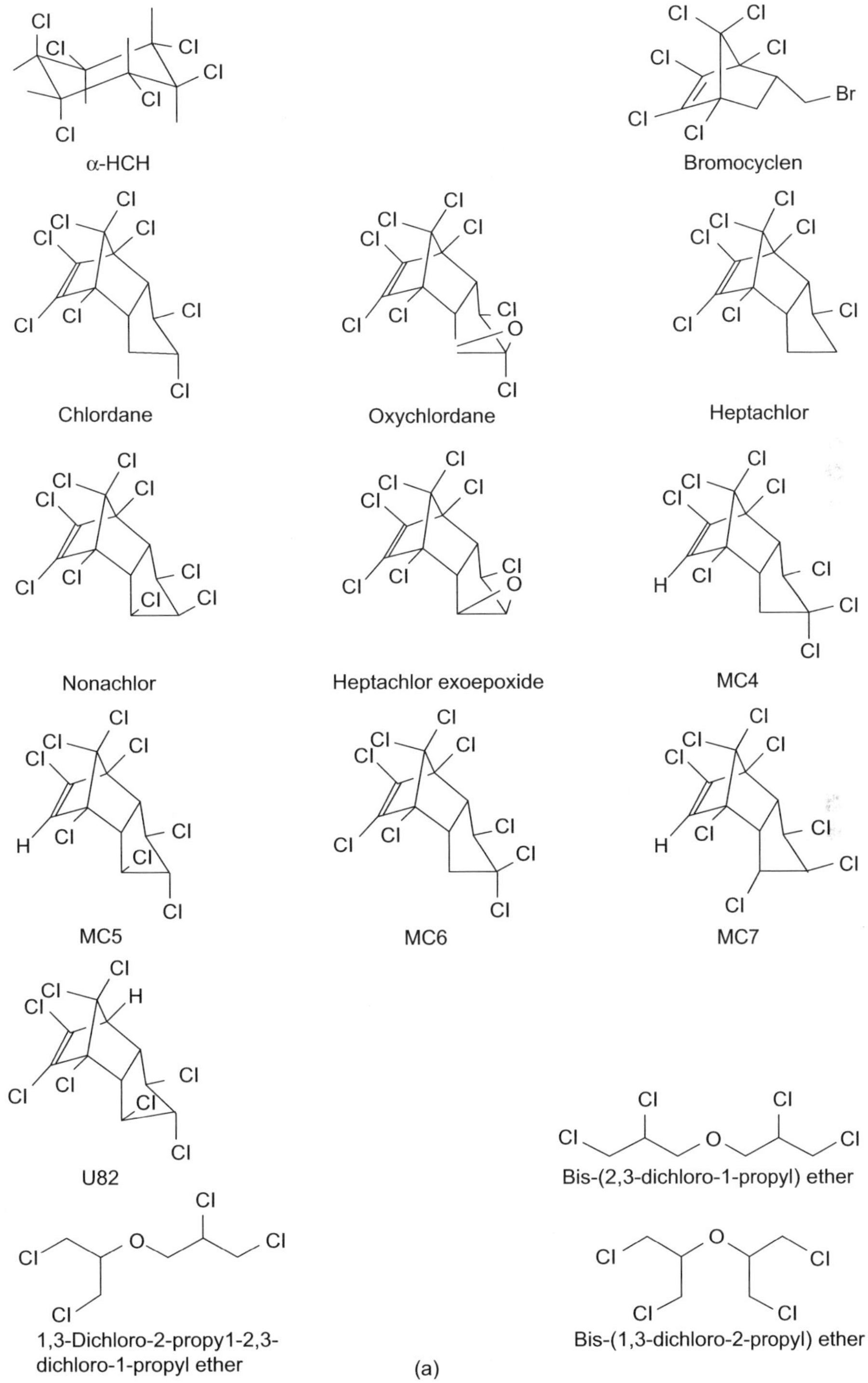

FIGURE 1-4 The chemical structures of some organic pollutants. (a) Aliphatic organochlorine pesticides, (b) aromatic organochlorine pesticides, and (c) phosphorous pesticides.

air pollution can be an issue to ponder still at home. Studies have shown that the air in our homes can be even more polluted than outdoor air. Many Americans spend up to 90% of their time indoors, often at home. Therefore, breathing clean indoor air can have an important impact on health. People who are inside a great deal may be at greater risk of developing health problems, or having problems made worse by indoor air pollutants. These people include infants, young children, the elderly, and those with chronic illnesses. Some common indoor biological pollutants include animal dander (minute scales from hair, feathers, or skin), dust mite and cockroach parts, fungi (molds), infectious agents (bacteria or viruses), and pollen.

All of us are exposed to biological pollutants. However, the effects on our health depend upon the type and amount of biological pollution and the individual person. Some people do not experience health reactions from certain biological pollutants, while others may experience one or more of the reactions such as allergic, infectious, and toxic. Except for the spread of infections, allergic reactions may be the most common health problem with indoor air quality in homes. Allergic reactions can range from mildly uncomfortable to life-threatening, as in a severe asthma attack.

1.5 Methods of Water Treatment

The ever-growing population, unplanned urbanization, rapid industrialization, and unskilled utilization of natural water resources have led to the destruction of water quality in many parts of the world. In many developing countries, groundwater provides drinking water for more than one-half of the nation's population and is the sole source of drinking water for many rural communities and some large cities. However, due to industrial, agricultural, and domestic activities, a variety of chemicals can pass through the soil and potentially contaminate natural water resources and reservoirs. In recent years, various toxic chemicals/compounds have been widely detected at dangerous levels in drinking water in many parts of the world posing a variety of serious health risks to human beings. Keeping the view of serious health problems to human beings due to elevated concentration of toxic pollutants, there is an urgent need for developing cost effective and environmentally friendly processes to remove them from drinking water and to safeguard the health of affected citizens. A number of technologies are available with varying degree of success to control water pollution [43] which is discussed in detail as follows.

1.5.1 Adsorption

Among the various techniques of water treatment, the adsorption process is an attractive approach for water treatment, particularly if the adsorbent is cheap and does not require a pretreatment step before its application [44]. For many applications, this process has been proven to be superior to other techniques for a variety of reasons [45–47] including the simplicity of design, low cost, high removal efficiency, ease of operation, and availability. Further, this process can remove/minimize different type of pollutants and thus, it has a wider applicability in water pollution control. Activated carbon is undoubtedly considered as universal adsorbent for effluent treatment and is commonly used for the removal of various pollutants from water [48]. However, its widespread use in wastewater treatment is sometimes restricted due to its higher cost. A large variety of nonconventional adsorbents have been examined for their ability to remove various types of pollutants from water and wastewater and have been reviewed extensively [49–59]. Activated carbon as widely used as the adsorbent material [60] as very low concentrations can be achieved and high loadings of pollutant on these adsorbents are possible. Adsorption processes are normally operated using a batch of adsorbent with sufficient capacity to operate for many months before reaching saturation. Once loaded the adsorbent must be disposed of or regenerated. Analysis of the whole life costs of water treatment by adsorption indicates that most of the treatment costs are associated with regeneration [61]. In spite of this, most studies of adsorption have focused on the development of adsorbents with high capacity and very few on developing adsorbents that can be easily regenerated. It has been found that various low-cost adsorbents developed from different origins show little or poor sorption potential for the removal of aquatic pollutants as compared to commercial activated carbon [62–65] regenerate activated carbon adsorbents electrochemically. However, due to its relatively low conductivity, an electrolyte was required, the time and energy required for regeneration was prohibitive, and the degree of regeneration was often significantly less than 100%. Brown and Roberts [66] have shown that graphite intercalation compounds can be used to remove dissolved organic compounds from water and that these adsorbents can be rapidly and fully regenerated by electrochemical oxidation. Meaghan and Gagnon [67] conducted in batch and column adsorption experiments using a groundwater from Halifax Regional Municipality that had an average arsenic concentration of 43 mg/L and a pH of 8.1. The residual solids (As) in this study were from five water treatment plants, four surface water treatment plants that utilized either alum, ferric, or lime in their treatment systems, and one iron removal plant. In batch adsorption experiments, iron-based residual solids and lime-based residual solids performed similarly to GFH, a commercially available adsorbent, while alum-based residual solids performed poorly. Langmuir isotherm modeling showed that ferric residuals had the highest adsorptive capacity for arsenic (Q_{max} ¼ 2230 mg/kg and 42,910 mg/kg), followed by GFH (Q_{max} ¼ 640 mg/kg), lime (Q_{max} ¼ 160 mg/kg), and alum (Q_{max} ¼ <1 and 3 mg/kg). Similarly, the maximum arsenic removal was $>93\%$ for the ferric and lime residuals and GFH, while the maximum arsenic removal was $<49\%$ for the alum residuals under the same conditions. In a column adsorption experiment, ferric residual solids achieved arsenic removal of $>26{,}000$ bed volumes before breakthrough past 10.0 mg As/L, whereas the effluent arsenic concentration from the GFH column was under the method detection limit at 28,000 bed volumes. Overall, ferric and lime water treatment residuals (WTRs) were promising adsorbents for arsenic adsorption from the groundwater, and alum WTRs did not achieve high levels of arsenic adsorption. Vinitnantharat et al. [68] reports the use of a pellet of adsorbent made from water treatment sludge (S) and acid activated water treatment sludge (SH) for removal of fluoride in the batch equilibration technique. The influence of pH, adsorbent dosage, temperature, and effect of other ions were employed to find out the feasibility of acid activated adsorbent to remove fluoride to the permissible concentration of 0.7 mg/L. The results from the adsorption isotherm followed both Langmuir and Freundlich models and the highest fluoride removal was found for adsorbent activated with acetic acid at 2.0 mol/L. The optimum adsorbent dosage was found at 40 g/L, 0.01 mol/L acid activated adsorbent which was able to absorb fluoride from 10 down to 0.11 mg/L. The adsorption capacity was decreased when the temperature increased. This revealed that the adsorption of fluoride on SH was exothermic. In the presence of nitrate and carbonate ions in the aqueous solution, fluoride removal efficiency of SH decreased from 94.4% to 86.6% and 90.8%, respectively. However, there is no significant effect in the presence of sulfate and chloride ions. Ha et al. [69] studied the effect of chlorine on adsorption because prechlorination is often applied in order to control microorganisms and taste-and-odor-causing materials, which may influence organics removal by adsorption and membrane filtration. Thus, the addition of chlorine into an advanced water treatment process using a hybrid of adsorption and ultrafiltration (UF) was investigated in terms of natural organic matter (NOM) removal and membrane permeability. A comparison between two adsorbents, iron oxide particles (IOPs) and powdered activated carbon, was made to understand the sorption behavior for NOM with and without chlorination. Chlorine modified the properties of dissolved and colloidal NOM in raw water, which brought about lower total organic carbon (TOC) removal, during IOP/UF. The location of IOPs, whether they were in suspension or in a cake layer, affected NOM removal, depending on the presence of colloidal particles in feed water. Chlorine also played a role in reducing the size of particulate matter in raw water, which could be in close association with a decline in permeate flux after chlorination. Guo et al. [70] reported that carbon-doped silicon oxide (SiCOH) dielectrics are one of the most suitable candidates for advanced low dielectric constant (low-k) interlayer material. To improve water adsorption resistance, the plasma enhanced chemical vapor deposited SiCOH films have been posttreated by the NH_3 plasma for various times, and the resulting SiCOH films are thus examined by water adsorption experiments. In his study, the results indicate that the SiCOH films treated by the NH_3 plasma exhibit enhanced resistance against water adsorption. Further, Fourier-transform infrared spectroscopy and X-ray photoelectron spectroscopy are used to characterize the chemical composition and bonding states of the pristine and NH_3 plasma-treated SiCOH films. It is revealed that the plasma surface treatment leads to the formation of Si–N, C(sp3)–N, C(sp2)-N, (N–)n Si–C ($n = 1$–3) configurations and loss of carbon atoms. Ippolito et al. [71] studied the aluminum-based WTRs can adsorb waterborne and soilborne P, As(V), As(III), and perchlorate and may be able to adsorb excess environmental selenium. WTR, clay minerals, and amorphous aluminum hydroxide were shaken for 24 h in selenate or selenite solutions at pH values of 5–9 and then analyzed for selenium content. Selenate and selenite adsorption edges were unaffected across the pH range studied. Selenate adsorbed on to WTR, reference mineral phases, and amorphous aluminum hydroxide occurred as outer sphere complexes (relatively loosely bound), while selenite adsorption was identified as inner-sphere complexation (relatively tightly bound). Selenite sorption to WTR in an anoxic environment reduced Se (IV) to Se(0), and oxidation of Se(0) or Se(IV) appeared irreversible once sorbed to WTR. Al-based WTR could play a favorable role in sequestering excess Se in affected water sources. Wibulswas et al. [72] developed a method for the removal of phenolic compounds from water by surfactant-modified pillared clays. An organophilic pillared clay-based adsorbent was prepared by incorporating a cationic surfactant, hex-adecyltrimethyl ammonium (HDTMA) during the synthesis of conventional aluminum pillared clays. A specific amount of HDTMA was desired to occupy the cation exchange sites of the precursor clay. The presence of the HDTMA surfactant enhanced the adsorption ability of the clays toward phenolic compounds from aqueous systems. Adsorption experiments were carried out using a batch equilibration technique to investigate the removal of three toxic phenol compounds—phenol, 3-monochlorophenol, and 3,5-dichlorophenol by the surfactant-modified pillared clays. Comparative adsorption experiments were also conducted with montmorillonite, the starting clay, as a reference material. The experimental results demonstrate the feasibility of utilizing surfactant-modified pillared clays as an adsorbent for removing phenol and chlorinated phenols from water. The adsorption affinity decreased in the following sequence: 3,5-dichlorophenol > 3-monochlorophenol > phenol. The Langmuir model was used to analyze the adsorption equilibria and calculate the adsorption capacity of phenolic compounds on these surfactant-modified pillared clays, S-PILCs.

1.5.2 Reverse Osmosis

With the rapidly increasing demands on water resources, freshwater shortage has become an important issue affecting the economic and social development in many countries. As one of the main technologies for producing freshwater from saline water and other wastewater sources, reverse osmosis (RO) has been widely used so far. RO is a pressure-driven process whereby a semipermeable membrane (i.e. RO membrane) rejects dissolved constituents in the feeding water but allows water to pass through [73] the RO technology therein, which has been developed more than half a century of industrial operation, is considered as a promising way and gaining worldwide acceptance at present [74–76]. Al-Rifai et al. [77] developed a RO method for the removal of pharmaceuticals and endocrine disrupting compounds in a water recycling process. The author examined the concentrations of 11 pharmaceuticals from various therapeutic categories and two endocrine disrupting chemicals by full-scale microfiltration (MF) and RO membrane facilities. Salicylic acid was the most abundant analyte with a concentration range of 11–38 μg/L, followed by bisphenol A with concentrations ranging from 6 to 23 μg/L. The concentration of all analytes decreased on average by one order of magnitude following primary and secondary treatment. Gemfibrozil, primidone, and carbamazepine were found to have lower removal efficiencies (74–78%) than other compounds during these stages, which could indicate lower biodegradability. The MF and RO systems were found to further lower the pollutant concentrations by an order of magnitude. The overall removal efficiencies in the final recycled water were above 97%, resulting in product water concentrations of lower than 0.1 μg/L for most compounds. An exception to this finding was observed for bisphenol A, which was detected in concentrations up to 0.5 μg/L in the final recycled water. Den et al. [78] reported that desalination of seawater and brackish water by RO has become increasingly important for drinking water supply in a greater part of the world. The presence of high silica concentrations in some brackish water, however, limits the application of RO desalination due to the potential formation of silica scales that irreversibly deteriorate the membrane material and performance. The author investigates the feasibility of electrocoagulation as a pretreatment process to remove silica from the source brackish water. The author found that bipolar configuration attained greater extent of silica removal as compared to monopolar configuration. Increases in charge loading generally improved the silica removal efficiency, but excessive hydraulic retention time (HRT, 60 min) was detrimental to the system performance. In this study, with no modification of the source water, silica removal efficiency up to 80% was achieved under a current intensity of 0.5 A and a HRT of 30 min. The subsequent nanofiltration (NF) studies demonstrated severe flux declines over the first 3 h, yielding only 70% of its initial flux for brackish water containing 100 mg/L silica and progressively lower with higher silica concentrations. For the pretreated water by electrocoagulation, the extent of flux decline was markedly improved, suggesting that the pretreatment was effective for the attenuation of the flux decline. Öner et al. [79] developed a cross-flow flat sheet RO method for the removal of both the boron and silica from geothermal water. The authors were used four different types of RO membranes, AD-SWRO (GE Osmonics), AG-BWRO (GE Osmonics), BW-30-BWRO (FILMTEC), and AK-BWRO (GE Osmonics) to compare their performances. In order to examine the effects of pressure and cross-flow velocity, system was operated at pressures ranging from 15 to 30 bar and cross-flow velocities of 0.16 and 0.33 m/s. Obtained permeate boron concentrations were evaluated according to drinking and irrigation water standards. Senthilmurugan et al. [80] separated the inorganic and organic compounds by using a radial flow hollow-fiber RO module. Phenol, 2,4-dinitrophenol (DNP), pentachlorophenol (PCP), NaCl, NaBr, and KBr were separated from aqueous solutions under a variety of operating conditions by using a Permasep B9 hollow-fiber RO module. The permeate characteristics for all compounds were analyzed by the combined film Spiegler–Kedem (CFSK) model available in the literature. The errors between experimental and theoretical predictions were less than 15% for both permeate flow rates as well as permeate concentrations. The model parameters such as the solute permeability and the reflection coefficient were constant for all phenolic compounds and KBr at constant temperature and pH. However, in the case of NaCl and NaBr, the solute permeability increased as the feed concentration increased to 15,000 ppm, after which the concentration of the solute permeabilities became independent of the feed concentration. On the other hand, the reflection coefficient for both NaCl and NaBr went through a minimum and then increased to a maximum before reaching a constant value as the feed concentration was increased. It is also shown that for both organic and inorganic compounds. The CFSK model provides much better theoretical predictions than the combined film solution diffusion model.

1.5.3 Membrane Filtration

The presence of free chlorine content that is used as a disinfectant in conventional treatment is found to react with residual NOM. This reaction process has been found to have a tendency to form DBPs such as trihalomethanes, haloacetic acids (HAAs), and other halogenated organics. DBPs are carcinogens, and direct exposure can lead to cancers, miscarriages, and nervous system complications. Its small size, easier maintenance, and superior water quality produced by membrane filtration have made this advanced technology possible to replace conventional treatment processes that consist of ozonation, precipitation, coagulation, flocculation, chlorination, and gravel filtration [81]. In addition, the membrane filtration process offers the extra advantages over conventional treatment such as a small footprint, compact module, lower energy consumption, environmental friendliness, and the capability of handling wide fluctuations in feed quality. Membrane filtration processes involving MF, UF, and NF in potable water production have increased rapidly over the past decade. MF and UF are employed to remove microparticles and macromolecules, which generally include inorganic particles, organic colloids (i.e. microorganisms), and dissolved organic matter. However, the use of membrane processes does not directly eliminate the problem of DBPs; even though MF, UF, NF, or RO has excellent performance on the removal of microbial particles, the disinfection process is still necessary. Recently, Byun et al. [82] developed a Mn oxide-coated membrane and compared it with the Ti and Fe and found that the performance of the manganese oxide coated membrane was superior to that of the other membranes tested, showing the fastest recovery in permeate flux when ozone was applied and the greatest reduction in the TOC in the permeate. The removal of trihalomethanes (THMs) and HAAs precursors using the membrane coated 20 times with manganese oxide nanoparticles was significantly better than that for the membranes coated with 30 or 40 times with manganese oxide nanoparticles or 40 times with iron oxide nanoparticles. Fuchs et al. [83] developed a membrane filtration method for the treatment of water which is coupled with stirred tank reactor. By using a cross-flow membrane bioreactor, high anaerobic conversion rates of three different types of wastewater with varying organic content were achieved. Loading rates obtained were as follows: 20 g chemical oxygen demand (COD)/L/day for artificial wastewater, approximately 8 g COD/L/day from vegetable processing industry (sauerkraut brine), and 6–8 g COD/L/day for wastewater from an animal slaughterhouse. At stable conditions, COD removal rates in all three wastewaters were higher than 90%. Methane yields from the treatment of artificial wastewater, sauerkraut brine, and animal slaughterhouse wastewater were in the range of 0.17–0.30, 0.20–0.34, and 0:1220:32 L/ng/COD fed, respectively. The complete retention of biomass and suspended solids is a unique feature of this treatment process, which combines a high loading capacity and at the same time, high COD removal rates even for complex wastewater containing high concentrations of particulate matter. Moslemi et al. [84] described the bromated formation in a hybrid ozonation ceramic membrane filtration. The effect of pH, ozone mass injection rate, initial bromide concentration, and membrane molecular weight cut off (MWCO) on bromate formation in a hybrid membrane filtration ozonation reactor was studied. Decreasing the pH, significantly reduced bromate formation. Bromate formation increased with increasing gaseous ozone mass injection rate, due to increase in dissolved ozone concentrations. Greater initial bromide concentrations resulted in higher bromate concentrations. An increase in the bromate concentration was observed by reducing MWCO, which resulted in a concomitant increase in the retention time in the system. A model to estimate the rate of bromate formation was developed. Good correlation between the model simulation and the experimental data was achieved.

1.5.4 Electrochemical

Over the last 20 years, considerable attention has been focused on the elimination of pollutants through reactions activated by nontraditional methods. Electrochemical methods such as direct oxidation, indirect oxidation, and electro-Fenton (EF) have been demonstrated to be effective on a number of organic pollutants dissolved in water. In addition, direct reduction and metal deposition have been proved to be viable techniques for the elimination of nitrates and heavy metals from aqueous solutions. In this area, the literature mainly focus on fundamental research: considerable results have been achieved in terms of detailed scientific knowledge of the reaction mechanisms, but very few data are available on the behavior of laboratory-scale reactors; furthermore, the energetic balance of the overall process has rarely been analyzed. De Francesco et al. [85] developed a of electrochemical reactors for the treatment of polluted water and the result of their study was that there is a striking difference between the various electrochemical processes in terms of the electrode surface required to purify a certain flow rate of wastewater with a certain level of pollutant concentration. In this respect, direct oxidation appears the most promising route in view of future applications. In spite of this, an evaluation of the energy necessary to run the process shows that all the methods of electrochemical abatement require costs which appear higher than with non-electrochemical abatement techniques. Randazzo et al. [86] analyzed the two chlorinated aliphatic hydrocarbons by the electrochemical treatments. Acidic aqueous solutions of the chlorinated aliphatic hydrocarbons 1,2dichloroethane (DCA) and 1,1,2,2tetrachloroethane (TCA) have been treated by the EF process. Bulk electrolyses were performed at constant current using a boron-doped diamond (BDD) anode and an air diffusion cathode able to generate H_2O_2 in situ, which reacts with added Fe^{2+} to yield $^{\bullet}OH$ from Fenton's reaction. At 300 mA, almost total mineralization was achieved at 420 min for solutions containing 4 mM of either DCA or TCA. Comparative treatments without Fe^{2+} (anodic oxidation, AO) or with a Pt anode led to a poorer mineralization. The better performance of the EF process with BDD is explained by the synergistic action of the oxidizing radicals, BDD(•OH) at the anode surface and •OH in the bulk, and the minimization of diffusional limitations. The decay of the initial pollutant accomplished with pseudo-first-order kinetics. Chloroacetic and dichloroacetic acids were the major by-products during the degradation of DCA and TCA, respectively. Acetic, oxalic, and formic acids were also identified. The proposed reaction pathways include oxidative and reductive (cathodic) dechlorination steps. Chlorine was released as Cl^-, being further oxidized to ClO_3^- and, mostly, to ClO_4^-, due to the action of the largely generated BDD(•OH). Palma-Goyes et al. [87] explored the applicability of electrochemical oxidation on a triphenylmethane dye compound model, hexamethylpararosaniline chloride (or crystal violet, CV), using BDD anodes. The effect of the important electrochemical parameters: current density (2.5–15 mA/cm^2), dye concentration (33–600 mg/L), sodium sulfate concentration (7.1–50.0 g/L), and initial pH (3–11) on the efficiency of the electrochemical process was evaluated. The results indicated that while the current density was lower than the limiting current density, no side products (hydrogen peroxide, peroxodisulfate, ozone, and chlorinated oxidizing compounds) were generated and the degradation, through •OH radical attack, occurred with high efficiency. Analysis of intermediates using GC–MS investigation identified several products: *N*-methylaniline, *N,N*-dimethylaniline, 4-methyl-*N,N*-dimethylaniline, 4-methyl-*N*-methylaniline, 4-dimethylaminophenol, 4-dimethylaminobenzoic acid, etc. The presence of these aromatic structures showed that the main CV degradation pathway is related to the reaction of CV with the •OH radical. Under optimal conditions, practically 100% of the initial substrate and COD were eliminated in approximately 35 min of electrolysis, indicating that the early CV by-products were completely degraded by the electrochemical system. Sirés et al. [88] first time reported the electrochemical degradation of β-blockers, which are pharmaceutical pollutants with a high occurrence in natural waters. The oxidation ability of two preeminent, eco-friendly electrochemical advanced oxidation processes, namely AO and EF, has been compared at lab-scale by carrying out bulk electrolyses at pH 3.0 at constant current using a carbon-felt cathode able to electrogenerate H_2O_2 in situ. The studies of single component aqueous solutions were focused on atenolol as a model β-blocker. The AO process was proven much more effective using a large surface area BDD anode than a Pt one, which was explained by the great amount of active hydroxyl radicals BDD(•OH) and the minimization of their parasitic reactions. The EF process with a Pt anode and 0.2 mmol/L Fe^{2+}showed even higher performance, with fast destruction of atenolol following pseudo-first-order kinetics and fast mineralization because the oxidation process in the bulk allows overcoming the mass transport limitations. The time course of the concentration of the aromatic and short-chain carboxylic acid intermediates demonstrated the progressive detoxification of the solutions. Almost 100% of the initial N content was accumulated as NH^{4+}. Multicomponent solutions containing atenolol, metoprolol, and propranolol, which usually occur together in the aquatic environment, were treated by EF using the Pt/carbon-felt cell. A high mineralization rate was observed up to the overall TOC removal, which allowed reducing the energy consumption. The absolute rate constant for the reaction of each β-blocker with •OH was determined and the reactivity was found to increase in the order: atenolol (1.42×10^9 L/mol/s) < metoprolol (2.07×10^9 L/mol/s) < propranolol (3.36×10^9 L/mol/s).

1.5.5 Biological

A method of treating wastewater is described. Firstly, providing specially acclimatized microbial consortia with relatively low sensitivity to changes in pH and changes in temperature, such that the microbial consortia is capable of metabolizing contaminants efficiently within a predetermined broad pH range and broad temperature range. Secondly, maintaining the microbial consortia in a space limited and a substrate limited mode within an immobilized bioreactor operating within the predetermined pH range and the predetermined temperature range. Thirdly, introduce wastewater with liquefied insoluble organic contaminants into the immobilized bioreactor. The organic contaminants are metabolized by the microbial consortia working under space limited and substrate limited conditions, thereby treating the wastewater with a minimal sludge production rate. The groundwater treatment containing Mn(II) and Fe(II) has been investigated by Katsoyiannis et al. [89]. The biological oxidation of Mn(II) and Fe(II) in upflow filtration units comprised the applied experimental technique. The oxidation processes were mediated by specific bacteria, namely the *Leptothrix ochracea* and *Gallionella ferruginea,* which belong to the general category of manganese and iron oxidizing bacteria. The author focused on the characterization of the products of biological oxidation and on the examination of the kinetics of Mn(II) removal as compared with Fe(II) removal from groundwater. The products of biological oxidation were characterized using the spectroscopic techniques XRD, XPS, and SEM-EDS and comprised a mixture of biogenic hydrous manganese and iron oxides. The oxidation state of manganese in the precipitates was found to be between 3 and 4. Iron oxides were mainly in the form of amorphous ferrihydrite. The kinetic results indicated that the rates of manganese and iron oxidation were several orders of magnitude greater than the respective for abiotic oxidation. The bacterially mediated oxidation of iron was faster than manganese oxidation, presenting half-lives of reaction 0.9 and 3.98 min, respectively. Shawaqfeh [90] investigated the biodegradability of wastewater containing priority pollutant pesticide [Vydine or triadimenol ($C_{14}H_{18}CLN_3O_2$)] in different bioreactor configurations. The examined two laboratory-scale biological reactors: one reactor under aerobic condition and the other under anaerobic condition. The aerobic reactor was operated at an ambient temperature [(22 ± 2) °C], while the anaerobic reactor was run in the lower mesophilic range [(30 ± 2) °C]. The effect of pesticide concentration, HRT, and co-substrate on the treatment process was explored, using glucose as a supplemental carbon substrate. More than 96% pesticide was removed after an acclimation period of approximately 172 (aerobic) and 230 days (anaerobic). The aerobic reactor achieved complete Vydine utilization at feed concentrations up to 25 mg/L. On the other hand, the anaerobic reactor was able to degrade 25 mg/L of Vydine. Moreover, glucose was consumed first throughout the experiment in a sequential utilization pattern. The combination of anaerobic and aerobic biological processes yielded higher biomass concentration and lower retention time than individual units. The biomass in the combined reactors was first acclimated with the corresponding pesticide. Then, the target pesticide, at a concentration of 25 mg/L, was sequentially treated in a semi batch mode in the reactors. HRT studies showed that 24 h HRT of aerobic and 12 h HRT of anaerobic were the optimum combination for the treatment of simulated wastewater containing Vydine, which produced Vydine effluent. Abou-Elela et al. [91] developed a method for saline wastewater treatment using a salt-tolerant microorganism. A salt-tolerant microorganism (*Staphylococcus xylosus*) was isolated from a vegetable pickled plant containing about 7.2% salt. The isolated microorganism was used as inoculum for biodegradation. An activated sludge reactor operated in a fed-batch mode was used for the treatment of synthetic saline wastewater using three different microbial cultures namely activated sludge (100%), a mixture of *Staphylococcus* supplement by activated sludge (1:1), and pure *S. xylosus* (100%) at different salt concentrations ranging from 0.5% to 3% NaCl. The results obtained showed that at low NaCl concentration (1%), the removal efficiency of COD using different microbial cultures were almost the same (80–90%). However, increasing the NaCl concentration to 2% and using *Staphylococcus*-supplemented mixture by activated sludge and *S. xylosus* alone improved the treatment performance as indicated by COD removal rates which reached 91% and 93.4%, respectively, while the system performance started to deteriorate when activated bacterial culture was used alone (74%). Furthermore, the increase in NaCl concentration up to 3% and with the inclusion of *Staphylococcus*-supplemented mixture by activated sludge increased the COD removal to 93%, while the use of *S. xylosus* alone further improved the COD removal rate up to 94%. Also, the use of *S. xylosus* alone proved to be capable for biological treatment of a real case study of a vegetable pickled wastewater containing 7.2% salinity; the removal efficiency of COD reached 88% at this very high concentration of NaCl.

1.5.6 Water Treatment In Situ

The increasing need for properly treated water aimed for human consumption and for recycling purposes has motivated a substantial research effort in the last decade. In this context, the so called advanced oxidation processes (AOPs) have been identified as an attractive option for water treatment, particularly in cases where the contaminant species are difficult to remove using the usual biological or physicochemical technologies [92]. Specific examples of potentially important technologies of this type include seminal works that report the successful use of AOPs to treat aqueous solutions containing pesticides, dichloroactetic acid, phenol, and other organo-halogenated compounds [93–95]. Among the most promising AOPs for water contaminated by organic molecules, application of the Fenton reagent (an aqueous mixture of Fe^{2+} and H_2O_2 that produces hydroxyl radicals) stands out due to its high oxidation power and rapid oxidation kinetics

[96]. The main problems associated with its widespread use, however, involve the high cost of commercial H_2O_2 and the operational costs associated with handling and maintaining the proper concentration of this compound during the water-purification process. In recent years, however, the in situ electrochemical generation of the Fenton reagent, also known as the EF approach, has been studied in order to exploit the fact that the reduction of dissolved oxygen to peroxide can be carried out selectively in acidic medium on the surface of a carbonaceous cathode [97–103]. Silyn-Roberts and Lewis [104] studied the presence of the ammonia-oxidizing group *Nitrosomonas* in developing biofilms from a subsurface flow dairy effluent treatment wetland was investigated using in situ hybridization and a crude semiquantitative PCR. Fluorescently labeled oligonucleotide probes, specific for the *Nitrosomonas europaea* and *Nitrosomonas eutropha* species, showed that nitrifying bacteria were present from the onset of biofilm development but that over a 5-month period, the proportions remained low (averaging less than 1% of the total bacterial population). The greatest presence of these nitrifiers was found in the second trench of the subsurface flow wetland. *Nitrosomonas* were also found to be susceptible to environmental change: adverse environmental conditions, such as low temperatures, coincided with negligible *Nitrosomonas* presence throughout the wetland. However, during one period of elevated ammonium concentrations, *Nitrosomonas* presence rapidly increased as high as 7% of the total bacterial population. Investigations of relative occurrence of effluent bacterial population of *N.* europaea and the biofilm populations showed that *Nitrosomonas* presence in the effluent coincided with significant presence in adjacent biofilms. This suggests seeding between both sessile and planktonic populations of this species. *Nitrosomonas* occurred as single cells in biofilms throughout the study and were never observed as dense clusters. Li et al. [105] developed fully automated analytical method for the online in situ analysis of selected semi-VOCs in water. The method used a large-volume injection/gas chromatography/mass spectrometry coupled with a fully automated microscale solid-phase extraction technique, which was based on *x–y–z* robotic techniques. Water samples were extracted by using a 96-well solid-phase extraction plate. For most analytes included in this study, the obtained linear calibrations ranged from 0.05 to 5.0 μg/L with correlation coefficients of 0.996–1.000, the method detection limits were less than 0.1 μg/L, and the relative recoveries were in the range of 70–120% with a relative standard deviation of less than 15% for fortified reagent water samples. The applications to chlorinated tap water, well water, and river water have been validated. The obtained results were similar to those resulting from fortified reagent water samples for all analytes except metribuzin, bromacil, aldrin, and methoxychlor. Matrix effects were observed for these analytes. In general, this fully automated analytical method was rugged, reliable, and easy to operate and was capable of providing real-time data to water treatment and distribution systems as well as water reservation and protection systems. In addition, the method could reduce the analytical costs associated with sample collection, transportation, storage, and preparation.

References

[1] L.B. Franklin, Wastewater Engineering: Treatment, Disposal and Reuse, McGraw Hill, Inc., New York, 1991.

[2] R.L. Droste, Theory and Practice of Water and Wastewater Treatment, John Wiley & Sons, Inc., New York, 1997.

[3] D.Z. John, Hand Book of Drinking Water Quality: Standards and Controls, Van Nostrand Reinhold, New York, 1990.

[4] http://www.nrcan.gc.ca.

[5] J.H. Feth, Chloride in Natural Continental Water—A Review, U.S. Geological Survey Water-Supply Paper 2176, 1981, p. 30.

[6] S. Qi, Z. Zhang, Z. Zheng, Q. Gong, Y. Zhang, Saltwater intrusion in the Laizhou Gulf, Shandong Province, China: causes and its impact on coastal areas, Ambio 36 (4) (2007 June) 361–362.

[7] National Atmospheric Deposition Program, National Atmospheric Deposition Program 2000 Annual Summary, National Atmospheric Deposition Program Data Report 2001–01, 2001, p. 16.

[8] R.A. Krieger, J.L. Hatchett, J.L. Poole, Preliminary Survey of the Saline-Water Resources of the United States, U. S. Geological Survey Water-Supply Paper 1374, 1957, p. 172.

[9] Task Committee on Saltwater Intrusion, Saltwater intrusion in the United States, J. Hydraulics Div. 95 (HY5) (1969) 1651–1669. American Society of Civil Engineers.

[10] J.H. Feth, et al., Preliminary Map of the Conterminous United States Showing Depth to and Quality of Shallowest Ground Water Containing More Than 1,000 Parts per Million Dissolved Solids, U.S. Geological Survey Hydrologic Investigations Atlas, 1965.

[11] E. Custodio, Detection, in: Seawater Intrusion in Coastal Aquifers—Guidelines for Study, Monitoring and Control, Food and Agriculture Organization of the United Nations Water Reports 11, Rome, Italy, 1997, pp. 6–22 (Chapter 2).

[12] B.F. Jones, A. Vengosh, E. Rosenthal, Y. Yechieli, Geochemical investigations, in: J. Bear, et al. (Eds.), Seawater Intrusion in Coastal Aquifers—Concepts, Methods and Practices, Kluwer Academic Publishers, Dordrecht, The Netherlands, 1999, pp. 51–71.

[13] F.T. Manheim, M.K. Horn, Composition of deeper subsurface waters along the Atlantic continental margin, Southeast. Geol. 9 (4) (1968) 215–236.

[14] H. Meisler, The Occurrence and Geochemistry of Salty Ground Water in the Northern Atlantic Coastal Plain, U.S. Geological Survey Professional Paper 1404–D, 1989, p. 51.

[15] B.C. Douglas, M.S. Kearney, S.P. Leatherman, Sea Level Rise—History and Consequences, Academic Press, San Diego, California, 2001, p. 232.

[16] A.S. Navoy, Aquifer–Estuary Interaction and Vulnerability of Ground-Water Supplies to Sea-Level-Rise Driven Saltwater Intrusion: Ph.D. Dissertation, State College, Pennsylvania State University, 1991, p. 225.

[17] A.S. Avoy, G.B. Carleton, Ground-Water Flow and Future Conditions in the Potomac–Raritan–Magothy Aquifer System, Camden Area, New Jersey, New Jersey Geological Survey Report GSR 38, 1995, p. 184.

[18] U.S. Environmental Protection Agency, Secondary Drinking Water Regulations—Guidance for Nuisance Chemicals, U.S. Environmental Protection Agency Report EPA 810/K–92–001, at URL: http://www.epa.gov/safewater/consumer/2ndstandards.html, 1992 (accessed 22.09.2000).

[19] D.S. Powars, The Effects of the Chesapeake Bay Impact Crater on the Geologic Framework and the Correlation of Hydrogeologic Units of Southeastern Virginia, South of the James River, U.S. Geological Survey Professional Paper 1622, 2000, p. 53.

[20] T.E. Reilly, Analysis of ground-water systems in freshwater-saltwater environments, in: W.M. Alley (Ed.), Regional Ground-Water Quality, Van Nostrand Reinhold, New York, 1993, pp. 443–469.

[21] C.D. Rail, Groundwater Contamination, , In: Contamination, Sources, & Hydrology, vol. 1, Technomic Publishing Company, Inc., Lancaster, Pennsylvania, 2000.

[22] Pollution – Definition from the Merriam-Webster Online Dictionary, Merriam-webster.com, 2010-08-13 (retrieved 26.08.2010).

[23] M.A. Khan, A.M. Ghouri, Environmental pollution: its effects on life and its remedies, Int. Refereed Res. J. II (2) (April 2011).

[24] H. Fereidoun, M.S. Nourddin, N.A. Rreza, A. Mohsen, R. Ahmad, H. Pouria, The effect of long-term exposure to particulate pollution on the lung function of Tehranian and Zanjanian students, Pak. J. Physiol. 3 (2) (2007) 1–5.

[25] L. Rosenstock, The environment as a cornerstone of public health, Environ. Health Perspect. 111 (7) (2003) A376–A377.

[26] D.E. Kromm, Response to air pollution in Ljubljana, Yugoslavia, Ann. Assoc. Am. Geogr. 63 (2) (1973) 208–217.

[27] D. Mara, S. Cairncross, Guidelines for Safe Use of Wastewater and Excreta in Agriculture and Aquaculture: Measures for Public Health Protection, World Health Organization, Geneva, 1989, p. 187.

[28] Government of Pakistan, Economic Survey of Pakistan, Finance Division, Economic Division Wing, Islamabad, 2009.

[29] L.M. Schell, M.V. Gallo, M. Denham, J. Ravenscroft, Effects of pollution on human growth and development: an introduction, J. Physiol. Anthropol. 25 (1) (2006) 103–112.

[30] F.W. Carter, Pollution problems in post-war Czechoslovakia, Trans. Inst. Br. Geogr. 10 (1) (1985) 17–44.

[31] H.E. Huppert, R.S.J. Sparks, Extreme natural hazards: population growth, globalization and environmental change, Philos. Trans. R. Soc. 364 (1845) (2006) 1875–1888.

[32] World Health Organization (WHO), Air Quality: Volcanic Ash Cloud over Europe. Retrieved from: http://www.euro.who.int/en/what-we-do/health-topics/environmental-health/airquality/volcanic-ash-cloud-over-europe, 2010.

[33] D.H. Pink, Investing in Tomorrow's Liquid Gold, April 19, 2006.

[34] L. West, World Water Day: A Billion People Worldwide Lack Safe Drinking Water, March 26, 2006.

[35] The National Water Quality Inventory: Report to Congress for the 2002 Reporting Cycle – A Profile, United States Environmental Protection Agency (EPA), Washington, DC, 2007. Fact Sheet No. EPA 841-F-07-003.

[36] G. Allen Burton Jr., R. Pitt, Stormwater Effects Handbook: A Toolbox for Watershed Managers, Scientists, and Engineers, CRC/Lewis Publishers, New York, 2001, ISBN 0-87371-924-7 (Chapter 2).

[37] L. Ritter, K.R. Solomon, J. Forget, M. Stemeroff, C. O'Leary, Persistent Organic Pollutants, United Nations Environment Programme, 1995 (retrieved 16.09.2007).

[38] The Dirty Dozen, United Nations Industrial Development Organization (retrieved 17.12.2007).

[39] D.H. Lee, I.K. Lee, K. Song, M. Steffes, W. Toscano, B.A. Baker, D.R. Jacobs Jr., A strong dose–response relation between serum concentrations of persistent organic pollutants and diabetes, Diabetes Care 29 (2006) 1638–1644. PMID 16801591 (retrieved 6.09.2007).

[40] http://www.epa.gov/iaq/biologic.html.

[41] http://publications.usa.gov/epublications/indoorair-hazards/mold.htm, http://www.epa.gov/iaq/biologic.html.

[42] http://bigpower1.com/biopoll.php.

[43] F.W. Pontius, Water Quality and Treatment, fourth ed.). McGraw-Hill Inc., New York, 1990.

[44] S.B. Wang, Y. Boyjoo, A. Choueib, Z.H. Zhu, Removal of dyes from an aqueous solution using fly ash and red mud, Water Res. 39 (1) (2005) 129–138.

[45] R. Sanghi, B. Bhattacharya, Review on decolorisation of aqueous dye solutions by low cost adsorbents, Color. Technol. 118 (5) (2002) 256–269.

[46] V. Meshko, L. Markovska, M. Mincheva, A.E. Rodrigues, Adsorption of basic dyes on granular activated carbon and natural zeolite, Water Res. 35 (14) (2001) 3357–3366.

[47] Y. Bulut, H. Aydin, A kinetics and thermodynamics study of methylene blue adsorption on wheat shells, Desalination 194 (2006) 259–267.

[48] R.C. Bansal, M. Goyal, Activated Carbon Adsorption, CRC Press, 2005.

[49] S.J.T. Pollard, G.D. Fowler, C.J. Sollars, R. Perry, Sci. Total Environ. 116 (1992) 31.

[50] G.S. Miguel, S.D. Lambert, N.J.D. Graham, J. Chem. Technol. Biotechnol. 81 (2006) 1685.

[51] D. Mohan, C.U. Pittman Jr., J. Hazard. Mater. B137 (2006) 762.

[52] D. Mohan, C.U. Pittman Jr., J. Hazard. Mater. 142 (2007) 1.

[53] M. Ahmaruzzaman, Adv. Colloid Interface Sci. 143 (2008) 48.

[54] W.S.W. Ngah, M.A.K.M. Hanafiah, Bioresour. Technol. 99 (2008) 3935.

[55] D. Sud, G. Mahajan, M.P. Kaur, Bioresour. Technol. 99 (2008) 6017.

[56] S.H. Lin, R.S. Juang, J. Environ. Manag. 90 (2009) 1336.

[57] A. Bhatnagar, M. Sillanpää, Chem. Eng. J. 157 (2010) 277.

[58] M. Rafatullah, M. Sulaiman, R. Hashim, A. Ahmad, J. Hazard. Mater. 177 (2010) 70. K.Y. Foo, B.H. Hameed, Adv. Colloid Interface Sci. 152 (2009) 39.

[59] A. Srinivasan, T. Viraraghavan, J. Environ. Manag. 91 (2010) 1915.

[60] G. McKay, M.S. Otterburn, J.A. Aga, Fullers earth and fired clay as adsorbents for dyestuffs – equilibrium and rate studies, Water Air Soil Pollut. 24 (3) (1985) 307–322.

[61] EPA. Technologies for Upgrading Existing or Designing New Drinking Water Treatment Facilities United States EPA Technology Transfer Report EPA/625/4e89/023, 1989.

[62] R.M. Narbaitz, Cen, Electrochemical regeneration of granular activated carbon, Water Res. 28 (8) (1994) 1771–1778.

[63] M. García-Oton, F. Montilla, M.A. Lillo-Ródenas, E. Morallón, J.L. Vazquez, Electrochemical regeneration of activated carbon saturated with toluene, J. Appl. Electrochem. 35 (3) (2005) 319–325.

[64] R.M. Narbaitz, A. Karimi-Jashni, Electrochemical regeneration of granular activated carbons loaded with phenol and natural organic matter, Environ. Technol. 30 (1) (2009) 27–36.

[65] M.H. Zhou, L.C. Lei, Electrochemical regeneration of activated carbon loaded with p-nitrophenol in a fluidized electrochemical reactor, Electrochim. Acta 51 (21) (2006) 4489–4496.

[66] N.W. Brown, E.P.L. Roberts, Electrochemical pre-treatment of effluents containing chlorinated compounds using an adsorbent, J. Appl. Electrochem. 37 (11) (2007) 1329–1335.

[67] M.K. Gibbons, G.A. Gagnon, Adsorption of arsenic from a Nova Scotia groundwater onto water treatment residual solids, Water Res. 44 (2010) 5740–5749.

[68] S. Vinitnantharat, S. Kositchaiyong, S. Chiarakorn, Removal of fluoride in aqueous solution by adsorption on acid activated water treatment sludge, Appl. Surf. Sci. 256 (2010) 5458–5462.

[69] T.-W. Ha, K.-H. Choo, S.-J. Choi, Effect of chlorine on adsorption/ultrafiltration treatment for removing natural organic matter in drinking water, J. Colloid Interface Sci. 274 (2004) 587–593.

[70] H.-W. Guo, L. Zhu, L. Zhang, S.-J. Ding, D.W. Zhang, R. Liu, Influence of NH_3 plasma treatment on chemical bonding and water adsorption of low-k SiCOH film, Microelectron. Eng. 85 (2008) 2114–2117.

[71] J.A. Ippolito, K.G. Scheckel, K.A. Barbarick, Selenium adsorption to aluminum-based water treatment residuals, J. Colloid Interface Sci. 338 (2009) 48–55.

[72] R. Wibulswas, D.A. White (Fellow), R. Rautiu, Adsorption of phenolic compounds from water by surfactant-modified pillared clays, Trans. IChemE 77 (Part B) (March 1999).

[73] L. Malaeb, G.M. Ayoub, Reverse osmosis technology for water treatment: state of the art review, Desalination 267 (2011) 1–8.

[74] M. Schiffler, Perspectives and challenges for desalination in the 21st century, Desalination 165 (2004) 1–9.

[75] C. Fritzmann, J. Löwenberg, T. Wintgens, T. Melin, State of-the-art of reverse osmosis desalination, Desalination 216 (2007) 1–76.

[76] L.F. Greenlee, D.F. Lawler, B.D. Freeman, B. Marrot, P. Moulinc, Reverse osmosis desalination: water sources, technology, and today's challenges, Water Res. 43 (2009) 2317–2348.

[77] J.H. Al-Rifai, H. Khabbazb, A.I. Schäferc, Removal of pharmaceuticals and endocrine disrupting compounds in a water recycling process using reverse osmosis systems, Sep. Purif. Technol. 77 (2011) 60–67.

[78] W. Den, C.-J. Wang, Removal of silica from brackish water by electrocoagulation pretreatment to prevent fouling of reverse osmosis membranes, Sep. Purif. Technol. 59 (2008) 318–325.

[79] S. Görkem Öner, N. Kabay, E. Güler, M. Kitis, M. Yüksel, A comparative study for the removal of boron and silica from geothermal water by cross-flow flat sheet reverse osmosis method, Desalination 283 (2011) 10–15.

[80] S. Senthilmurugan, S.K. Gupta, Separation of inorganic and organic compounds by using a radial flow hollow-fiber reverse osmosis module, Desalination 196 (2006) 221–236.

[81] M. Clever, F. Jordt, R. Knauf, N. Rabiger, R. Rudebusch, H. Scheibel, Process water production from river by ultrafiltration and reverse osmosis, Desalination 131 (2000) 325–336.

[82] S. Byun, S.H. Davies, A.L. Alpatova, L.M. Corneal, M.J. Baumann, V.V. Tarabara, S.J. Masten, Mn oxide coated catalytic membranes for a hybrid ozonation–membrane filtration: comparison of Ti, Fe and Mn oxide coated membranes for water quality, Water Res. 45 (2011) 163–170.

[83] W. Fuchs, H. Binder, G. Mavrias, R. Braun, Anaerobic treatment of wastewater with high organic content using a stirred tank reactor coupled with a membrane filtration unit, Water Res. 37 (2003) 902–908.

[84] M. Moslemi, S.H. Davies, S.J. Masten, Bromate formation in a hybrid ozonation–ceramic membrane filtration system, Water Res. 45 (2011) 5529–5534.

[85] M. De Francesco, P. Costamagna, On the design of electrochemical reactors for the treatment of polluted water, J. Clean. Prod. 12 (2004) 159–163.

[86] S. Randazzo, O. Scialdoneb, E. Brillas, I. Sirés, Comparative electrochemical treatments of two chlorinated aliphatic hydrocarbons. Time course of the main reaction by products, J. Hazard. Mater. 192 (2011) 1555–1564.

[87] R.E. Palma-Goyes, F.L. Guzmán-Duque, G. Peñuela, I. González, J.L. Nava, R.A. Torres-Palma, Electrochemical degradation of crystal violet with BDD electrodes: effect of electrochemical parameters and identification of organic by-products, Chemosphere 81 (2010) 26–32.

[88] I. Sirés, N. Oturan, M.A. Oturan, Electrochemical degradation of β-blockers. Studies on single and multicomponent synthetic aqueous solutions, Water Res. 44 (2010) 3109–3120.

[89] I.A. Katsoyiannis, A.I. Zouboulis, Biological treatment of Mn(II) and Fe(II) containing groundwater: kinetic considerations and product characterization, Water Res. 38 (2004) 1922–1932.

[90] A.T. Shawaqfeh, Removal of pesticides from water using anaerobic-aerobic biological treatment, Biotechnology and Bioengineering, Chin. J. Chem. Eng. 18 (4) (2010) 672–680.

[91] S.I. Abou-Elela, MM Kamel, M.E. Fawzy, Biological treatment of saline wastewater using a salt-tolerant microorganism, Desalination 250 (2010) 1–5.

[92] A. Wenzel, A. Gahr, R. Niessner, TOC-removal and degradation of pollutants in leachate using a thin-film photoreactor, Water Res. 33 (1999) 937–946.

[93] W. Gernjak Gernjak, T. Krutzler, A. Glaser, S. Malato, J. Caceres, R. Bauer, A.R. Fernández-Alba, Photo-Fenton treatment of water containing natural phenolic pollutants, Chemosphere 50 (2003) 71–78.

[94] S. Malato, J. Blanco, C. Richter, M.I. Maldonado, Optimization of pre-industrial solar photocatalytic mineralization of commercial pesticides: application to pesticide container recycling, Appl. Catal. B Environ. 25 (2000) 31–38.

[95] S. Malato, J. Blanco, A. Vidal, C. Richter, Photocatalysis with solar energy at a pilot-plant scale: an overview, Appl. Catal. B Environ. 37 (2002) 1–15.

[96] C. Walling, Fenton's reagent revisited, Acc. Chem. Res. 8 (1975) 125–131.

[97] E. Brillas, M.A. Banos, J.A. Garrido, Mineralization of herbicide 3,6-dichloro-2-methoxybenzoic acid in aqueous medium by anodic, electro-Fenton and photoelectro-Fenton, Electrochim. Acta 48 (2003) 1697–1705.

[98] B. Boye, M. Mariéme Dieng, E. Brillas, Electrochemical degradation of 2,4,5-trichlorophenoxyacetic acid in aqueous medium by peroxi-coagulation: effect of pH and UV light, Electrochim. Acta 48 (2003) 781–790.

[99] E. Guivarch, N. Oturan, M.A. Oturan, Removal of organophosphurus pesticides from water by electrogenerated Fenton's reagent, Environ. Chem. Lett. 1 (2003) 165–168.

[100] M.A. Oturan, An ecologically effective water treatment technique using electrochemically generated hydroxyl radicals for in situ destruction of organic pollutants: application to herbicide 2,4-D, J. Appl. Electrochem. 30 (2000) 475–482.

[101] M.A. Oturan, N. Oturan, C. Lahitte, S. Trevin, Production of hydroxyl radicals by electrochemically assisted Fenton's reagent: application to the mineralization of an organic micropollutant, pentachlorophenol, J. Electroanal. Chem. 507 (2001) 96–102.

[102] Z. Quiang, J.H. Chang, C.-P. Huang, Electrochemical generation of hydrogen peroxide from dissolved oxygen in acidic solutions, Water Res. 36 (2002) 85–94.

[103] A. Ventura, G. Jacquet, A. Bermond, V. Camel, Electrochemical generation of the Fenton's reagent: application to atrazine degradation, Water Res. 36 (2002) 3517–3522.

[104] G. Silyn-Roberts, G. Lewis, In situ analysis of *Nitrosomonas* spp. in wastewater treatment wetland biofilms, Water Res. 35 (1) (2001) 2731–2739.

[105] Y. Li, J.E. George, C.L. McCarty, Online in situ analysis of selected semi-volatile organic compounds in water by automated microscale solid-phase extraction with large-volume injection/gas chromatography/mass spectrometry, J. Chromatogr. A 1176 (2007) 223–230.

2

Water Treatment for Inorganic Pollutants by Adsorption Technology

CHAPTER OUTLINE

Environmental Water, http://dx.doi.org/10.1016/B978-0-444-59399-3.00002-7

2.1 Introduction

Water is an essential component of this universe and plays a vital role in the proper functioning of the Earth's ecosystems. In spite of this, drinking water is also an alarming problem for some parts of the world. The quality of water resources is deteriorating exponentially due to contamination. Due to the geometrical growth of world population, domestic and agricultural activities, modern industrialization and civilization, and other geological, environmental and global changes, water pollution is continuously increasing day by day. Nowadays, water pollution is a very serious issue as it is very harmful for all living creatures, recreation, household, transportation fishing, and other commercial activities [1,2]. The importance of water quality preservation and improvement is necessary in life and increasing continuously. In view of these facts, all water regulatory authorities and agencies are very conscious on the pollution of water resources. More than 600 organic and inorganic pollutants have been reported in water along with microbial populations. Among these pollutants, heavy metal ions are very dangerous due to their toxic and carcinogenic nature. Moreover, some metal ions are not biodegradable or bio-transformable and, hence, exist in the environment for long time. Elements from fourth period of the periodic table namely cadmium, platinum, mercury, arsenic, antimony etc are very toxic [3,4]. In view of these points, attempts have been made to review the removal of these toxic metal ions from wastewater.

Various methods for water purification and recycling have been developed and used from time to time [5,6]. These methods have been classified on the basis of physical, chemical, electrical, thermal and biological principles. These methods include screening, centrifugation and filtration, micro- and ultra filtration, sedimentation, crystallization, and gravity separation, precipitation, flotation, oxidation, evaporation, coagulation, solvent extraction, distillation, ion exchange, reverse osmosis, electrolysis, adsorption, electrodialysis, etc. Among these methods, ion exchange, reverse osmosis, electrolysis, electrodialysis, and desorption are very efficient technologies. The cost of water purification by these technologies, ranges from 10 to 450 US$ per cubic meter of treated water but the cost of water treatment using by adsorption techniques ranges 5 to 200 US$ per cubic meter of water. Therefore, adsorption technique for the treatment of water is the best method due to its wide range of applications, efficient and ease of operation. This technique is also considered as universal water treatment and reclamation technique as it can be applicable for the removal of organics, inorganic and biological pollutants which are soluble and insoluble in water, and have a potential of 99% removal efficiency. On the other hand, pollutants at industrial level are removed from wastewater by using suitable adsorbents. Adsorption technique can also be studied for reclamation for potable, source reduction, industrial and other water purposes. Due to wide range of applications of this technique, a lot of work has been carried out on the removal of different pollutants and, hence, many authors have published articles on this issue [7,8]. In this article, author reviewed the applications of adsorption for the removal of metal ions from wastewater using adsorption technique.

2.2 Adsorption Essentials

The deposition of a particular component at the surface or interface between two phases is known as adsorption. Basically adsorption is a surface phenomenon. In this process, there are two components one is adsorbent, i.e. the compound (pollutant) that sticks or gets attached to the solid surface is called as adsorbate and the compound on which adsorption occurs is called as adsorbent. The adsorption capacity of the adsorbent may be change with the change of temperature, nature of adsorbates and adsorbents, presence of other pollutants and other atmospheric and experimental condition such as pH, temperature, concentration of pollutants, and contact time and particle size of the adsorbent. The presence of suspended particles, oils and greases also affect the efficiency of the process and, therefore, sometimes prefiltration is required. When an adsorbent is shaken with the contaminated/polluted water, the pollutants adhere to the surface of the adsorbent and an equilibrium is established. At this stage, the amount of pollutants adsorbed and in wastewater becomes constant. The relationship, at given temperature, between the equilibrium amount of the pollutants adsorbed and in wastewater is called the Langmuir, Freundlich adsorption isotherm, and other models are well-known adsorption models by which the adsorption efficiency of the pollutants can be explained in a systematic and scientific way [9]. These models are not described here and the interested readers should consult various textbooks on this subject.

2.3 Design of Batch and Column Adsorption Experiments

The design of a batch adsorption process starts with plotting the isotherm (Figure 2-1). The loading (mg) amount of material adsorbed per gram of adsorbent is plotted versus the concentration of the solution at equilibrium. The designing of a column for the adsorption starts with laboratory testing to determine the breakthrough capacity and mass transfer zone is also formed in the column bed by passing the contaminated water.

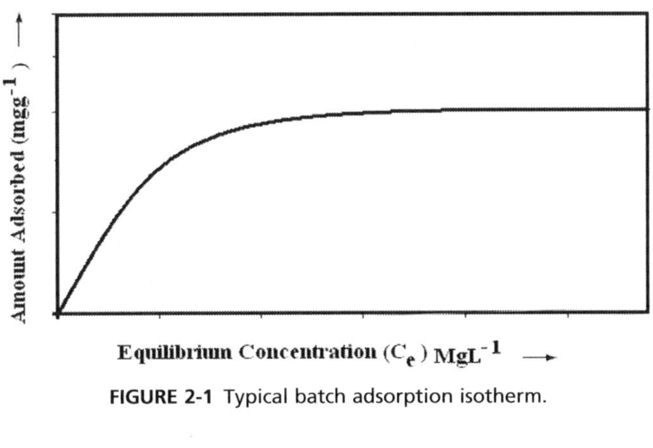

FIGURE 2-1 Typical batch adsorption isotherm.

The depth of this zone is maintained by the characteristics of the adsorbent and metal ions and hydraulic factors. The depth of mass transfer zone is a measure of physical and chemical resistance to mass transfer. This zone reaches up to the bottom of the column where metal ion concentration in effluent and becomes equal to the influent concentration and a breakthrough point of the column occurs. This condition is shown in Figure 2-2 and the breakthrough curve is idealized by the assumption that the removal of the solute is completed over the initial stages of operation. The break point is chosen arbitrarily at some low value, C_b for the effluent concentration. At an arbitrarily selected effluent concentration, C_x closely approaching to C_0, the sorbent is considered to be essentially exhausted. The total mass quantity of effluent V_b, passing per unit cross-section at the break point and the nature of the breakthrough curve between the values of V_b and V_x are used for design purpose. The primary adsorption zone in the fixed bed adsorbent is that part of the bed where there is a concentration reduction from C_x to C_b. It is assumed to be of constant length (Lm).

The fractional capacity (f) of the adsorbent at break point may be calculated as follows:

$$f = \int_0^1 \left[1 - \frac{C}{C_0}\right] d\frac{(V_e - V_b)}{(V_x - V_b)} \tag{2.1}$$

where C, C_0 and V_e are equilibrium concentration, initial concentration and volume of effluent at exhaustion.

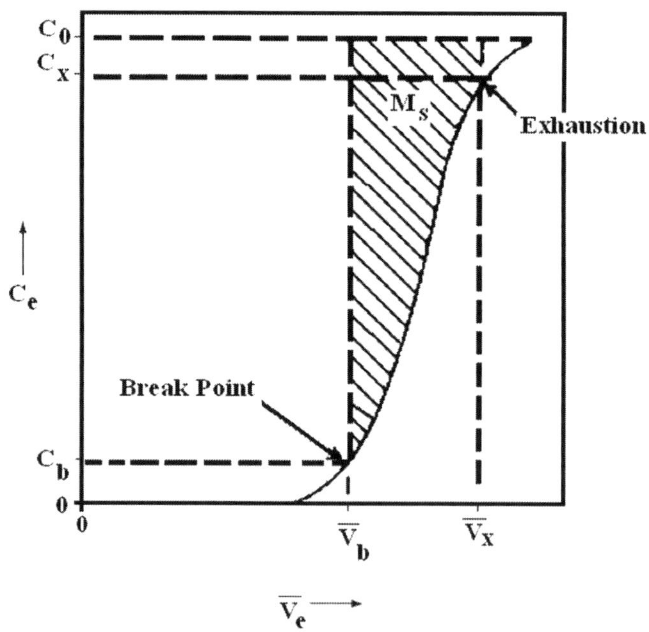

FIGURE 2-2 A typical column breakthrough curve. (C_o = initial concentration of influent, C_x = concentration of effluent at exhaustion, C_b = concentration of effluent at breakpoint, C_e = equilibrium concentration effluent, V_b = volume of effluent at breakpoint, V_x = volume of effluent at exhaustion, V_e = volume of total effluent).

The percent column saturation at break point is given by the following equation.

$$\%\text{Saturation} = [D + Lm(f - 1)]/D \quad (2.2)$$

where D is the adsorbent bed depth.

2.4 Activated Carbon Adsorbents

In 1940's, activated carbon was introduced for the first time as the main water industry's standard adsorbent for the reclamation of municipal and industrial wastewater to a potable water quality [10,11]. The use of carbon adsorption for the direct treatment of liquid industrial waste streams is a relatively recent practice but has been successfully utilized for the reclamation of metals from electroplating wastewater [12]. It has been found as a good adsorbent due to its high capacity of adsorption because of small particle sizes and active free valences. In spite of this, activated carbon could not be used as the adsorbent for water treatment at large scale due to its high cost of production. Moreover, the regeneration of activated carbon is difficult due to the use of costly chemicals and, hence, its regeneration is not easily possible at commercial scale.

2.5 Development of Low-Cost Adsorbents

These drawbacks associated with activated carbon have generated the interest for the development of its low-cost alternatives using a range of carbonaceous and other precursors [7,13]. There are so many starting materials for these replacements are from agricultural or industrial by-products and, hence, their use as secondary adsorbents contributes to waste minimization with recovery and reuse. Many types of low-cost adsorbents have been developed and used for wastewater treatment. We have categorized them into two group's, i.e. prepared from i) organic and ii) inorganic precursors; used for water purification (removal of metal ions).

2.5.1 Preparation of Adsorbents from Organic Precursors

The generally used organic precursors are bark, scrap tyres, tannin-rich materials, petroleum wastes, sawdust, rice hulls, coconut shell, fruit stones, peat moss, fertilizer wastes, algae, chitosan, seafood processing wastes, sea weeds, resins, wastes of some industries and fly ash. Some, inorganic precursors include hydroxides, metal oxides, red mud, clays, blast furnace slag, zeolites, soil, sediment, ore minerals etc.

2.5.1.1 Low-Cost Sources to Obtain Activated Carbon

Some wastes are also converted into activated carbons by some physico-chemical processes. Activated carbon includes a very high range of carbonaceous materials, which show a high degree of porosity and extended inter particle surface area [14]. The activated carbons contain high surface area of 600–2000 m^2/g and well-defined micro-porous structure of average pore opening of 1.5 nm [15]. Activated carbon is prepared by following two stage of operation and involves carbonization of the raw material below 800 °C in the absence of oxygen followed by the activation at elevated temperatures using an activate (steam, carbon dioxide or air) or sometimes chemical catalysts. The time for carbonization, heating rate, and activation are the most important factors, which control the final pore structure, surface area and the chemistry of the carbon prepared [16,17].

Different varieties of carbon-containing feedstocks have been used for the formation of activated carbon of various grades preferably by the catalytic activation of an initially pyrolyzed char. Generally (pine, 130,000 tones per year), anthracite and bituminous coal (100,000 tone per year) lignite (50,000 tone per year), peat (50,000 tone per year) and coconut shell (35,000 tone per year) are used as a feedstocks for the production of activated carbon at commercial scale. Some other alternative feedstocks such as olive stones and almond shells account for ~10,000 tone per year [8,14]. The carbon contents of these varieties have ranged from 50 to 90% (w/w) with a bulk density range of 0.40–1.45 g cm^{-3}. Basically, commercial carbon is prepared from pinewood and coal but still there is a continuous search for more inexpensive alternatives [7].

2.5.1.2 Carbonization and Activation

Carbonization and activation are the most important processes in the development of low-cost activated carbon. At the time of carbonization, pyrolytic decomposition of precursor occurs together with the concurrent elimination of many noncarbon species (H, N, O and S) [18]. In the process of carbonization, low molecular weight volatiles are first released followed by light aromatics and, finally, hydrogen gas [19], thus resulting in a fixed carbonaceous char [20]. In the process of carbonization, the pores are filled with tarry pyrolysis residues and, hence, activation is required in order to increase the internal surface of the material. Activation may be performed via a chemical or physical process. In the process of chemical activation, a catalyst is impregnated into the feedstock. The most commonly used chemicals are $ZnCl_2$, H_3PO_4, H_2SO_4, KOH, K_2S and KCNS for the activation [18]. A saturated solution of catalyst-impregnated feedstock is dried to influence pyrolysis in such a way that tar formation and volatilization can be kept at minimum. Two different temperatures are used for the activation of carbonized product [21], these temperatures are 400 °C followed by activation raising the temperature up to 800 °C as used for the industrially wood-based carbons.

The pore size distribution of the final product totally depends upon the degree of initial impregnation [15]. Kadlec et al. [22] studied the chemically and physically activated carbons and observed that the different mechanisms of activation are responsible for different shapes and sizes. The porosity in the activated carbon develops due to the increased burn-off of disorganized carbon [17]. Most common activants are steam, CO_2 and air for the mechanisms of these reactions [23,24]. A large number of agricultural and industrial wastes have high concentration of carbon content and, hence, shows a significant potential for the preparation of carbonaceous charcoals, which may be further activated to obtain porous adsorbents. The different waste products used for generating low-cost adsorbents are given in Table 2-1. The different types of waste precursors are discussed below.

2.5.1.3 Scrap Tyres

The management of waste tyres and their disposal are very important as they have serious fire risk. The combustion of this waste may lead to the release of large volumes of oil, phenols, polynuclear aromatic hydrocarbons and heavy metals into the atmosphere [25]. Contrarily, tyres contain high carbon content and, therefore, controlled pyrolysis was employed to convert waste tyres to semiactive carbon. The carbon content has been recovered by stripped tyres by moving bed reactor, at 400–700 °C, and a variety of atmospheres and the surface area of the activated carbon was in the range of 300 to 350 m^2/g [26]. A wet thermal process (water/steam 700–900 °C) was employed for vapor

Table 2-1 Waste Products used for Generating Low-Cost Adsorbents

Sl. No	Waste Materials
1	House hold wastes
	a. *Fruit waste.*
	b. *Coconut shell.*
	c. *Scrap tyres.*
2	Industrial waste
	a. *Petroleum wastes.*
	b. *Fertilizer wastes.*
	c. *Fly ash.*
	d. *Sugar industry wastes.*
	e. *Blast furnace slag.*
3	Agricultural waste
	a. *Bark and other tannin-rich materials.*
	b. *Sawdust and other wood type materials.*
	c. *Rice husk.*
	d. *Other agricultural waste.*
4	Soil and ore materials
	a. *Clays.*
	b. *Red mud.*
	c. *Zeolites.*
	d. *Sediment and soil.*
	e. *Ore minerals.*
5	Sea materials
	a. *Chitosan and seafood processing wastes.*
	b. *Sea weed and algae.*
	c. *Peat moss.*
	d. *Miscellaneous waste.*

phase applications with a highly active carbon of surface area 1260 m^2/g. Waste tyres were used for another powdered activated carbon with a lower surface area of 193 m^2/g. It exhibited a considerable affinity for many pollutants [27]. Due to costly operations involved in their production, these precursors could not achieve a good reputation in the development of low-cost charcoal alternative.

2.5.1.4 Sawdust and Lignin

Sawdust and other wood materials have been used as the alternatives of activated carbon. The adsorption capacities of these adsorbents are mainly due to the presence of lignin or tannin components rather than the cellulose content [28]. Sabadell and Krack [29] also found several wood types for the adsorption of Pb(II), Cd(II), Cu(II) and Ni(II) metal ions. These wood materials were converted by heat into low-cost activated carbons. The average capacities for the woods used in these studies are 0.0721 meq/g for ash shavings, 0.0982 meq/g for oak sawdust, and 0.0683 meq/g for cedar sawdust. Vaishya and Prasad [30] evaluated that the rate of adsorption of cupric ions on sawdust was very high initially and maximum adsorption occurs within 1 h. Zarraa has investigated the ability of sawdust to adsorb Cr (VI) from waste solutions in a batch reactor under forced convection conditions [31].

It was confirmed that the removal of Cr (VI) depend on the weight of sawdust. Ajmal et al. treated sawdust of mango tree (*Mangifera indica*) in 0.1 M disodium hydrogen phosphate for a period of 24 h followed by washing with distilled water and then dying at 40 °C [32]. A comparative study was performed by the same group, on the sorption capabilities of this phosphate treated and untreated sawdust, for the adsorption of Cr(VI). It was found that nearly 100% adsorption occurred below pH 2.0 for the initial Cr(VI) concentration of 8–50 mg/L. The experiments, carried out with synthetic as well as actual electroplating industry wastewaters containing 50 mg/L Cr(VI) showed 100% adsorption. For regeneration purpose, 87% desorption of Cr(VI) from phosphate-treated sawdust was achieved using 0.01 M NaOH. Raji and co-workers [33] studied the factors which influenced the uptake of Hg(II), Pb(II) and Cd(II) onto the activated carbon prepared from bicarbonate-treated wood sawdust and found that the percentage of adsorption increases with decrease in initial adsorbate concentration and increase in adsorbent concentration and temperature. It was also found that the maximum adsorption was in the range of pH 4–8 for Pb(II) and Hg(II) and 5–9 for Cd(II). These adsorbent can be reactivated by treating with 0.2 M HCl. A waste product of paper industry (black liquor) was used for the extraction of lignin and converted into a low-cost adsorbent for the removal of lead and zinc [34]. As it was found that the adsorption capacity at 30 °C is 1587.0 and 73.0 mg/g for Pb and Zn respectively, which increased to 1865 and 95 mg/g, respectively at 40 °C. Masri et al. [35] studied the adsorption capacity of Hg (II) and reported that the adsorption capacity is 150 mg/g. It is confirmed that the adsorption capacities of sawdust, lignin and other wood-type materials were high with short equilibrium time. The cost of the process development and regeneration of these adsorbents were very high. Moreover, the raw materials were not easily available at commercial level therefore this material cannot be used frequently.

The presence of sulfur-containing groups on the adsorbents increase the affinity for the adsorption of heavy metals but shows a low affinity for lighter metals. Some sulfur-containing compounds contain thiols, dithiocarbamates, sulfides, xanthates and dithiophosphates. Among these compounds, xanthates are highly insoluble and important precursors because they are easy to prepare with relatively inexpensive reagents [36]. The removal of heavy metal from xanthates is done by ion exchange reaction method. To make these effective as adsorbents, some xanthate sludge was made by solidification [37]. It was also observed that the adsorption capacity varied randomly due to variation in particle porosity and size. The decomposition and air oxidation of sulfur atoms decrease the sorption capacity of xanthates with time (18% in 47 days). The different adsorption capacity metals such as Cd(II) ranged from 0.28 to 0.38 meq/g [15.7–21.4 mg/g of Cd(II)] and for Mn(II), Co(II), Ni(II), Ag(II), Zn(II), Hg(II) and Pb(II) ranged 0.3 to 0.4 meq/g. Different types of starch xanthate, soluble (SSX) and insoluble (ISX) were compared for their adsorption capacities [38,39]. ISX starch xanthate were produced by cornstarch through an extra cross-linking step. It was also confirmed that ISX exhibited better properties than SSX in terms of metal removal capacities, reliabilities and ease of operation, while, SSX is more economical than ISX. The cost of ISX is US$ about $60 for 100 g and requires 90–110 h for preparation while the cost for SSX is only $10/100 g taking only 3–4 h for preparation. A preliminary estimated cost for ISX production was only $0.60/kg [40]. Wing [41] reviewed several case studies for demonstrating ISX effectiveness and found ISX was used to remove heavy metals from plating rinse waters. The adsorption principle is ion exchange, which limits ease of their regeneration due to high cost. It is also noted that the probabilities of decomposition and air oxidation of these adsorbents are serious drawbacks. The adsorption capacities of xanthates are not quite good; especially for light metal ions and, hence, they could not be used at pilot scale for wastewater treatment.

2.5.1.5 Bark and Tannin

Solid waste of bark timber can be used directly as a possible adsorbent due to its high percentage of tannin content. The active species of the tannin content are found to be polyhydroxy polyphenol groups in adsorption process. The uptake of cations takes place through ion exchange by displacing the adjacent phenolic hydroxyl groups forming a chelate [42,43]. The development of coloration and toxicity in water is the main problem due to soluble phenols associated with tannin-containing materials. To overcome this problem, chemical pretreatment of bark has been tried by a number of workers. Alves et al. [44] performed the formaldehyde pretreatment to diminish the bleeding of colored compounds without appreciably affecting capacity. Randall et al. [42,45] compared of the bark adsorption to that of peanut skins, walnut expeller meal and coconut husks and found their adsorption capacities to be comparable. Furthermore, Randall et al. [46] also tried formaldehyde pretreatment of peanut skins to prevent leaching of color and disintegration upon prolonged contact with water. The resulting product by this process was quite stable in water and was efficient for uptake of Cu(II), Hg(II), Cd(II), and Pb(II). Orhan and Buyukgungor [47] studied the adsorption power of tannin-containing products obtained from nut and walnut shell, waste tea and coffee and compared it with that of activated carbon. They concluded that the capacities of the tannin-containing products are only slightly less than that of activated carbon.

Edgehill and Lu [48] demonstrated that the potential of carbonized slash pine bark as a suitable substitute for activated carbon. The carbonization of bark was performed by slow heating in nitrogen for 6.5 h at 670 °C and surface area, average micropore and mesopore diameter, and micropore volume of the prepared adsorbent was 332 m^2/g, 2.17 A° and 0.125 cm^3/g, respectively. Al-Asheh and Duvanjak [49] studied the cadmium ions adsorption onto carbonized pine bark and concluded that the maximum cadmium uptake was obtained using very fine particles of pine bark. Guedes de Carvalho et al. found that Portuguese pine bark is a suitable precursor for activated carbon production [50]. The carbonization of ground bark was carried out at 600 °C for 1 h under nitrogen atmosphere and then activated between 800 and 1000 °C by partial CO_2 gasification. Aoyama et al. [51] evaluated the adsorption capacities of 34 conifer leaves with good adsorption capacities of *Ginkgo biloba, Taxus cuspidata, Cephalotaxus harringtonia, Taxodiaceae* and *Cupressaceae* species for Cr(III) metal ion.

The adsorption capacities of conifer leaves were compared with different commercial activated carbons and found that 3.12–5.09 mg/g Cr (III) was adsorbed on different conifer leaves while the adsorption capacities of activated carbons were 1.23–2.75 mg/g for Cr(III). The larch (*Larix leptolepis*) bark was able to remove Cr(VI) from dilute aqueous solutions at pH 3. Preparation of activated carbon is costly but bark and tannin-rich materials are having high adsorption capacities. Moreover, coloration and toxicity are developed in water. Therefore, they could not be used extensively for the removal of toxic metal ions from wastewater.

Patil et al. [52] carried out the experiment for the removal of nickel metal ions by using powder-activated charcoal and nonconventional adsorbents. Kalmykova et al. [53] used, sawdust, fiber sludge ash, pine bark and some other biological materials (peat, seaweed and shrimp shells) for adsorption of some metal ions from laboratory synthesized water and industrial landfill by using batch and column operations; with maximum adsorption occurring on peat. The addition of 10% by weight of fiber ash to the peat resulted into higher adsorption capacities of Cd(II), Ni(II) and Pb(II) but lower for Cu(II) and Zn(II), respectively. Subbaiah et al. [54] found biosorption of Ni(II) ions from aqueous solution by Acacia leucocephala bark by different pH, adsorbent dosage, metal ion concentration, temperature parameters and contact time. Khokhotva and Waara [55] performed the comparison of adsorption of Cu(II), Ni(II), Zn(II) and Pb(II) metal ions on plain and urea-treated pine bark of *Pinus sylvestris*.

2.5.1.6 *Rice Hulls*

Some other species like rice hulls or husks; agricultural wastes products; have been used for the production of low-cost adsorbents [rice husk carbon (RHC)], it is confirmed that adsorption efficiencies of these adsorbents are quite encouraging, i.e. RHC has shown 94.4% w/w Cd(II) adsorption capacity [56]. This efficiency was also compared with a commercial-activated carbon (filtrasorb 400) with 25.4% w/w removal capacity of Cd(II). However, the high cost of steam activation at 700 °C remains a major problem with RHC. Srinivasan et al. [57] performed the experiment with RHC for Cr(VI) removal from industrial wastewater and concluded that RHC was comparable to commercially available activated carbon. The column capacities for Cr(VI) uptake were found to be 8.9 and 6.3 mg/g of RHC and commercial carbon, respectively. The adsorbent was successfully used for the treatment of an actual wastewater from a real plating shop bearing Cr(VI) up to 300 mg/L. Roy et al. [58] carried out the study for the investigation of heavy metal uptake by rice hulls and observed that adsorption capacities of rice hulls were quite good for Cr(VI) and Pb(II). Mostafa [59] developed steam activated and zinc chloride activated carbons from rice husks and these activated carbon were used for the removal of Hg (II), Pb(II) and Cd(II) ions from aqueous solution at 298 °K. The uptake capacities of rice hulls were greater than commercially available activated carbons in both batch and column experiments.

2.5.1.7 *Paper Industry*

Guo et al. [60] carried out the adsorption of Cu(II), Cd(II), Zn(II), Pb(II), and Ni(II) on a lignin isolated from black liquor; a waste product of the paper industry. It was observed that adsorption was dependent on strong pH and ionic strength. The workers studied surface complexation modeling, which showed two main types of acid sites attributed to carboxylic- and phenolic surface groups responsible for adsorption; with maximum adsorption occurring by phenolic group. Similarly, removal of Cr(III) from waters through lignin was also done by Wu and co-workers [61], the effect of different pH, contact time, ionic strength, dosage and Cr(III) concentration were studied. It was confirmed by these studies that it followed pseudo-second-order and Langmuir models with a maximum of 17.97 mg/g. The authors reported ion-exchange mechanism for adsorption. They applied the developed method for removal of Cr(III) from real wastewater sample. Pejic et al. [62] studied waste short hemp fibers; after chemical removal of lignin and hemicelluloses; for adsorption of Zn(II), Cd(II) and Pb(II) metal ions from aqueous solutions. The effects of initial ion concentration, contact time and cosorption were studied in batch sorption experiments. The obtained results showed when the content of either lignin or hemicelluloses was progressively reduced by chemical treatment; the sorption properties of hemp fibers were improved. Short hemp fibers were capable of sorbing metal ions with maximum total uptake capacities for Pb(II), Cd(II) and Zn(II) 0.078 mM/g, and from ternary mixture 0.074, 0.035 and 0.035 mMl/g, respectively. Of course, the cost of processing was low in comparison to the ready-made-activated carbon and, definitely, these raw materials have quite good scope in wastewater treatment process.

2.5.1.8 Peat Moss

Peat moss is an inexpensive and easily available adsorbent material, with good adsorption capacities for a variety of pollutants. Peat is highly in demand as a fuel in Ireland and Northern Europe and as a soil conditioner in the United States [63]. Peat contains lignin and cellulose as the major constituents. Lignin has a large number of polar functional groups that can be employed for the chemical adsorption. Because of these properties, this variety tend to have a high cation adsorption capacity and can be useful for the removal of metal ions [64]. It is also well known that peat is highly porous (95%) and possess a large specific surface area (>200 m^2/g) [65]. Different types of peats are available with varying chemical and physical properties. Chen et al. [66] evaluated the complexation and ion-exchange facilitating adsorption of Cu(II) onto eutrophic and oligotrophic peat. It was determined that the efficiency of these adsorbents had been reported up to six times for Cr(VI) in comparison to activated carbon [67].

Kertman et al. [68] reported the efficacy of prepared peat to adsorb different heavy metals with the adsorption capacity of 90–230 mg/g for lead. It had been concluded that acid-treated peats have higher adsorption capacities than the normal ones for uptake of Cr (II) and SCd (II) [69]. Sharma and Forster reported the process for the regeneration of peat. Ho and Mckay [70] revaluated the kinetics of lead sorption on to peat. It was assumed that the batch sorption model, is based on pseudo-second-order mechanism, has been developed to predict the rate constant of sorption. Kalmykova et al. [71] studied the adsorption of Cu(II), Ni(II), Pb(II), Zn(II) and Cd(II) on Sphagnum peat. The rate constant were calculated for the second-order model and the order of removal capacity was Pb(II) > Cu(II) > Ni(II) > Cd(II) > Zn(II) with optimization by pH and ionic strength.

2.5.1.9 Fruit Stones

For the preparation of activated carbons, the outside or inner side hard parts of certain fruits such as olive stones, almond shells, apricot and peach stones, palm fruit bunch, coconut etc. are also reviewed as the prevailing raw materials for the preparation of activated carbons. The process for the formation of activated carbons follows the removal of the oily part of these raw materials by treatment under anaerobic conditions and subsequent drying and washing with organic solvents. The solid part is roasted in a rotating metal cylinder until its color changes from brown to black [72]. The cost of this developed material is less $50/ton in comparison to granulated-activated carbon having a $4500/ton cost. Rodriguez-Reinso and coworkers [73] carried out the optimization by using a variety of chemical and physical methods for the activation procedures for activated carbon production. Consequently, high-quality microporous carbons have been prepared and characterized having 90–1550 m^2/g surface area. Ferro-Garcia et al. [74] studied a series of fruit pit carbons for the removal of Cd(II), Zn(II) and Cu(II) from metal finishing plant wastewaters. Adsorption capacities for three metal ions ranged 18–150 μM/g. The prepared adsorbents were subsequently recommended by the authors for the treatment of metal finishing plant wastewaters.

Namasivayam and Periasamy [75] developed the procedure for the formation of activated carbon from bicarbonate-treated peanut hulls (BPHC) with 208 m^2/g surface area and investigated the adsorption of Hg(II). It was observed that quantitative removal of 20 mgd/m^3 Hg(II) in 100 mL by 70-mg adsorbent over a pH range of 3.5–10.0. it was found that that BPHC was 7.0 times more effective in comparison to that of GAC. Furthermore, Periasamy and Namasivayam [76] also studied the removal of lead from aqueous solutions using peanut hull carbon (PHC) and coal-based commercial granular activated carbon (GAC). Same research group used PHC adsorbent for the removal and recovery of cadmium from wastewater [77]. The adsorption capacity (Kf) of PHC was 31.0 times greater than that of GAC, as a result of a comparative study with a commercial GAC. Almost quantitative removal of 20 mg/L Cd(II) by 0.7 g of PHC/L of aqueous solution was observed in the pH range of 3.0–9.5. A range of chemically activated tamarind nut carbons was prepared by using a variety of treatments adapted from the carbon literature [78,79].

High-quality GACs were obtained by a well-known precursor, coconut shell and it is responsible for ~9% w/w global production of commercially activated carbon. It is also an inexpensive raw material due to its widespread distribution in the developing countries. Banerjee et al. [80] developed coconut shells-activated carbon by heating $ZnCl_2$ at 700 °C and resulting intermediate was activated in steam or air at 900 °C. In this process, the black-activated carbon had a surface area of 800 m^2/g with high adsorbing properties. Alaerts et al. [81] studied the removal efficiencies of commercial carbons having surface area similar to activated coconut-based carbon for the removal of aqueous Cr(VI) and found that coconut-based carbon was highly effective as 94% (w/w) of a 20 mg/L Cr(VI) solution in 4-h contact time during equilibrium batch studies. Mortley et al. [82] compared the characterized-activated carbon from materials of varying morphology including coconut husks and shells with commercially available carbons. The precursors with higher activation energies yielded higher surface areas and developed micro porosities on account of the lesser degree of lignocellulosic breakdown. Venezuelan coconut carbons were prepared by Laine and co-workers [83] prepared through chemical activation. The shells were impregnated with H_3PO_4 followed by a one-step carbonization/activation at 450 °C. The surface area of the product was 1200 m^2/g. The evaluation of the efficiency of copper impregnated coconut husk carbon (CuCHC) for the removal of arsenic from water and wastewater was done by Manju and coworkers [84].

The activated carbon of coconut husk carbon (CHC) was prepared by treating one part of coconut husk with 1.8 parts by weight of concentrated sulfuric acid (18 M) and heated with 150 °C for 24 h and then washed with distilled water to remove free acid and dried at 105 °C. It was also confirmed that CuCHC is 5 times more effective than CHC for the removal of As(III) and the maximum adsorption capacity of the developed adsorbent was 158 mg/g at 60 °C. The batch operation was also carried out for the removal of As(III) from actual industrial waste collected from one fertilizer industry in Cochin city (Kerala, India) using CuCHC. The results showed that the CuCHC can be used as an adsorbent for the effective removal of As(III) from water and wastewaters. Rao [85] studied the kinetics of Cr(VI) removal on prepared activated coconut shells. Different impregnation ratios of H_2SO_4, H_3PO_4 and $ZnCl_2$ activating agents were used for the preparation of activated carbons. It was found that the adsorption was maximum at pH 1.0 and increased with increase in impregnation ratio due to increase in surface area of coconut shell carbon and also concluded that the developed activated carbons from these raw materials were quite good having high activation energies, surface area and adsorbing properties. These adsorbents have been used under a wide range of experimental conditions with up to 31 times higher efficiencies than the commercially available activated carbon. But low abundance restricts their use at pilot scale.

2.5.1.10 Chitin and Chitosans

Chitin and chitosans are one of the most important materials for the removal of toxic metal ions due to their inexpensive and effective in nature. It is found in high concentration in the exoskeletons of crabs and other arthropods and also in the cell wall of some fungi [86,87]. The main source for chitin is crabmeat canning industry. More than 50,000 tons/annum of chitin is available from the fisheries of crustacean [88]. Chitosan or glucosamine is a derivative of chitin and can be prepared from chitin through deacylation. Chitosan is found in the cell wall of some fungi like *Mucorales* stains. About 2000 tons of chitosan could be produced at a cost of about $2.00 per kg [86]. With the help of deacetylation chitin generate free amino groups, which enhance the chelating capacity of chitosan to 5 to 6 times higher than that of chitin itself.

Chitosan showed greater adsorption capacities than that of poly-(*p*-aminostyrene) which are constituents of expensive ion-exchange resin. Tseng et al. [89] prepared the chitosan by using fishery wastes including shrimp shell, lobster shell, crab shell and cattle bone. Chitosan is successfully used for the removal of Cd(II), Cu(II) and Ni(II). Chui et al. [90] performed the procedure for the removal and recovery of Cu(II), Cr(III) and Ni(II) from solutions using crude shrimp chitin packed in small columns. With the help of dilute HCl, *Penaeus marginatus* were dematerialized in the form of shrimp shell wastes in small columns to yield shrimp chitin. It is reported that 96% calcium was efficiently removed by this method without removal of proteins. Wastewater of electroplating baths rinsing having Cu(II) gave encouraging results. The shrimp chitin removed 92–96% Cu(II) from copper pyrophosphate and acid copper bath rinsing water. Kim et al. [91] carried out the effect of deacetylation on sorption of chromium on chitin. Deacetylated chitins (10.7–67.2%) were prepared by alkaline hydrolysis on optimal degree of deacetylation, which could effectively removed chromium from textile effluent. Baran et al. [92] described adsorption of Cr(VI) ions onto various sorbents (chitin, chitosan, ion exchangers; Purolite CT-275 (Purolite I), Purolite MN-500 (Purolite II) and Amberlite XAD-7) in batch process. The optimizing parameters for the maximum efficiency were pH, agitation period and concentration of Cr(VI) ions. The optimum pH for Cr(VI) adsorption was found 3.0 for chitin and chitosan, whereas it did not affect adsorption on ion exchangers. The maximum chromium sorption occurred at 50, 40, 30 min, for chitin, Purolite II and Chitosan, Purolite I and Amberlite XAD-7, respectively. The workers demonstrated that adsorption could be increased by impregnating the reported adsorbents by ethylenediaminetetraacetic acid (EDTA). Over all, chitosan is a readily available, inexpensive adsorbent found suitable for removing maximum chromium from aqueous solution.

Jayakumar et al. [93] removed Ni(II), Zn(II) and Cu(II) metal ions by using alginate/phosphorylated chitin (p-chitin) blend films; prepared by mixing of 2% of alginate and p-chitin in water and then cross-linked with 4% $CaCl_2$ solution. With the result of these findings, it was confirmed that alginate/p-chitin blend films were bioactive. The optimizing parameters were pH, contact time and initial metal ion concentrations. The maximum adsorption capacities of alginate/p-chitin blend films for Ni(II), Zn(II) and Cu(II) at pH 5.0 were found to be 5.67, 2.85 and 11.7 mg/g, respectively. Briefly, abundance of these raw materials is quite high, i.e. available at large amount next to cellulose. Nevertheless, they could not achieve a good reputation of effective adsorbents in the removal of metal ions from wastewater because the regeneration of these adsorbent is required costly chemicals due to the strong chelation with metals.

2.5.1.11 Sea Weed

Seaweed can also be used for the removal of various pollutants from water. Vijayaraghavan and Yun [94] studied the adsorption of metal ions on various bioadsorbents such as fungi, bacteria, algae, and industrial and agricultural wastes. The properties of the cell wall constituents, such as peptidoglycan, and the role of functional groups, such as carboxyl, amine and phosphonate, were discussed on the basis of their biosorption properties.

One of the species of seaweed-like brown algae has significant ion-exchange properties associated with their polysaccharide contents. Brown marine algae, *Ascophyllum nodosum*, was used in sorption columns to remove Cd(II) with good adsorption capacity of approximately 67 mg/g [95]. Some single cell green algae such as *Fucus serratus* and *Laminaria digitata* showed better uptake of Cd(II) and Hg(II) [96]. It is found that the green seaweed *Ulva lactuca* is less effective than two brown seaweeds. The stability and mechanical properties were enhanced by the modifications of seaweeds by cross-linking. By cross-linking, the swelling of the material was reduced so that the material would be easily used for column adsorption. One of the algae, i.e. *A. nodosum* was found to adsorb Cd(II) (2.15 mg/g) while 149 mg/g was absorbed by formaldehyde cross-linked treated *A. nodosum*. It has demonstrated pronounced adsorption in comparison to other seaweeds experimented, i.e. *Sargassum natans* (135 mg/g) and *Fucus vesiculosus* (Cd(II) = 73 mg/g). Leusch et al. [97] performed the adsorption capacities of *A. nodosum* and *Sargassum fluitans* after formaldehyde cross-linking, glutaraldehyde crosslinking and polyethylenimine embedding for the removal of Cu(II), Ni(II), Cd(II), Pb(II) and Zn(II) metal ions. A high molecular weight polymer, algin was prepared by brown seaweed. It is a polysaccharide-based biosorbent and it can be prepared from algin by replacing protons in the carboxylic groups with metal ions. In the sorption process, the formation of alginate took place with the exchange of heavy metal ions with calcium ions. It is also demonstrated that the theoretical maximum binding capacity of sodium alginate is 138 mg/g for Cu(II) while a capacity of 107 mg/g was experimentally determined [98]. Araujo and Teixeira [99] developed the applicability of calcium alginate beads for the removal of Cr(III) from aqueous solutions.

The biomass of the marine alga *Sargassum baccularia* has been estimated for its adsorption and desorption of Cd(II) ions [100]. Desorption of Cd(II) was attained by desorbing agents such as HCl and EDTA using a batch reactor system. Both these desorbents were appropriate to remove the adsorbed Cd(II) from the biomass. It was found that HCl at pH 2.0 could desorb 80% of Cd(II) initially loaded on to the biomass. Almost complete recovery of Cd(II) was achieved by a 3.24 mM EDTA solution. Marine algal-based biosorbents for the removal of Pb(II) and Cu(II) was used by Malik et al. [101]. The weakly acidic carboxyl groups of polysaccharides present within the algal matrix showed high sorption capacity for both metals. Adsorption kinetics of Al(III), Zn(II), Hg(II), Pb(II), Cu(II), and Cd(II) onto living microalgae (*Scenedesmus subspicatus, Cyclotella cryptica* and *Phaeodactylum tricornutum*) has been examined [102]. The highest values for the rate constants obtained for *Porphyridium purpureum* followed by *P. tricornutum*. High values for the maximum content were attained for *C. cryptica* and *S. subspicatus*. The maximum rate constant was 24/h for the adsorption of Hg(II) on to *P. purpureum* whereas maximum metal content (0.243 g/g) was acquired for Zn on *C. cryptica*. Two strains of *Pseudomonas syringae* (Blue and Brown) to remove copper from aqueous solutions and compared with the synthetic Linde LZ-52Y aluminosilicate zeolite was used by Hall et al. [103]. The two bacterial strains were tolerant to copper and were able to grow in media with Cu(II) concentrations up to 1000 ppm. As in case of chitin and chitosan the basic process of metal ion adsorption was through ion-exchange mechanisms, which made the regeneration process complex and expensive. That is why costly HCl and EDTA chemicals were used to recover metal ions from these adsorbents. Moreover, disintegration and swelling tendencies of these biomasses constrained their use at large scale; especially in columns.

Equilibrium, thermodynamic and kinetic studies of Al(III) biosorption on brown algae (*Padina pavonica*) biomass with optimum conditions of pH, biomass dosage, contact time and temperature were described by Sari and Tuzen [104]. The biosorption capacity of *P. pavonica* biomass was found to be 77.3 mg/g. The metal ions were desorbed from *P. pavonica* using 1.0 M HCl. The high stability of *P. pavonica* allowed a slight decrease of about 20% in the recovery of Al(III) ions after 10 times of adsorption-elution process. The adsorption mechanism explained was chemical ion exchange with spontaneous and endothermic nature. Cd(II) metal ions on red macro alga (Hypnea valentiae) was removed by Rathinam et al. [105]. The optimized variables were cadmium concentrations, pHs and temperatures; with maximum adsorption at pH 6.0 (250 mg/L) were found to be optimized variable. The authors explained that the presence of neutral salts and other metal ions affected the cadmium uptake behavior of the biomass considerably. The biosorption characteristics of Ni(II) ions using brown algae (*Cystoseira indica, Nizmuddinia zanardini, Sargassum glaucescens* and *Padina australis*) by varying pH, contact time, initial metal concentration and temperature as experimental parameters was reported by Pahlavanzadeh et al. [106]. Esmaeili et al. [107] stated the removal of Cu(II) metal ions aqueous solution and wastewater on marine brown alga *Sargassum* species as adsorbent. The effect of pH, biosorption time, adsorbent dose and metal ion concentration were studied. The optimum pH was 4.0 with maximum removal of 94.83% within 2 h. The uptake of heavy metals by these adsorbents was reversible and the mechanisms of uptake were not well known. Therefore, the research on this issue is not entirely developed. Briefly, despite of reasonably good abundance, these materials could not be used at huge scale for the removal of metal ions. Therefore, the need of more extensive studies in terms of cost, performance, preparation methodologies, energy, efficiency, kinetics, regeneration and management is evident.

2.5.1.12 Biomass

The elimination of metal ions from wastewater can be performed utilizing the dead biomass of bacterial cell as appropriate adsorbents. The dead biomasses are effective due to the fact that the dead cells have better accumulation capacity for heavy metals than living cells. The presence of ionized groups in the cell wall, forcing the cell to attract metal cations makes the bacterial cell to have anionic surface. Huge amount of waste microbial biomass are formed in industries such as citric acid biosynthesis and penicillin production. About 790,000 tons of microbial waste are produced each year by fermentation industries, with 41,000 tons resulting from citric acid production by *Aspergillus niger* [86]. The problems of waste toxicity and nutrient requirements are removed using the dead rather than live biomass. Biomass are used for the elimination of Cd(II), [108] Pb(II), [109] etc. by many workers. The optimum pH range for biosorption of Pb(II) ion, on *Penicillium chrysogenum*, was 4 to 5. The saturated sorption of Pb(II) was 116 mg/g by dry biomass, which was higher than that of activated charcoal and some other microorganisms [108]. At pH 4.5 *P. chrysogenum* biomass exhibited a significant selectivity for Pb(II) compared to other metal ions such as Cd(II), Cu(II), Zn(II) and As(III). Sorption preference for metals increased in the order of As(III) > Zn(II) > Cu(II) > Cd(II) > Pb(II).

The removal of Pb(II) by live, resting and dead cells of a lignolytic white rot fungus, *Phanerochaete chrysosporium is studied by* Yetis et al. [110]. The fact that adsorption is two stage process, i.e. very rapid surface adsorption within the first hour and a slow intracellular diffusion after 2 h of metal exposure was revealed by kinetic studies. The results explained that the resting cells were able to uptake up to 80 mg/g Pb(II). The biomass originated from different growth phases exhibited different adsorption capacities for Pb(II). It showed that the older resting cells held lower Pb(II) adsorption capacities than the young ones. Bai and Abraham [111] studied the biosorption of Cr(VI) by chemically modified biomass of

Rhizopus nigricans and the possible mechanisms of Cr(III) complexation to the adsorbent. The cetyl trimethyl ammonium bromide (CTAB), polyethylenimine (PEI) and amino propyl trimethoxy silane (APTS) were utilized for conducting biomass modification experiments to enhance the biosorption efficiency up to exceptionally high levels. They were used for metal ions removal at laboratory scale regardless of anionic surface on the dead bacterial cells. Due to the selective nature of these cells with fair adsorption capacities, this material has limited use at pilot scale. The main drawback associated with regeneration process was the intracellular diffusion process in the dead bacterial cell. Sometimes, living bacterial cells may present along with dead cell, which contaminant wastewater with bacteria and this may be hazardous to health. In a nutshell, the dead bacterial biomass could not be used for the treatment of wastewater at large scale.

2.5.1.13 Fly Ash

Fly ash is an aggressive waste product discharged from thermal power plants, sugar industries, brick kilns and various other industries Fly ash has been tried as the suitable adsorbent for the removal of metal ions from water by several researchers. There is no set methodology to use fly ash and a large amount is dumped as land filler. The estimated global production of fly ash is 290 metric tons per annum [112] Fly ash is an inorganic residue and its chemical constituents comprise 40–50% (w/w) silica (SiO_2), 20–35% (w/w) alumina (Al_2O_3) and 5–12% (w/w) iron oxide (Fe_2O_3). Fly ash also contains 12-30% (w/w) carbon and unburnt residue. Many metal ions such as Cr(VI), Cu(II), Ni(II), Zn(II), Hg(II) etc. were removed from wastewaters [113,114]. Grover and Narayanaswamy [114] studied that fly ash can be exercised for the removal of Cr(VI). The adsorption capacity for Cr(VI) was pH depending and a maximum adsorption (4.25 mg/g) was achieved at pH 2.0 [115]. The adsorption capacity for copper, determined at pH 6.5, was 73–95% (w/w) [116]. Uptake was explained in terms of a chemisorptive surface complex between the negatively charged silica surface and the hydroxylated copper species in solution. These investigations revealed that fly ash wollastonite blend was most effective and reported 40–60 times cheaper than the activated carbon. Kapoor and Viraraghavan [117] explained that the properties of fly ash were extremely changeable and the adsorption capacity varied with the lime content. Furthermore, the same workers [118] studied the usefulness of fly ash in adsorbing mercury from wastewater. They found that a contact time of 2 h was necessary for the adsorption to reach equilibrium at pH between 5.0 and 5.5. Viraraghavan and Dronamraju [119] evaluated the effectiveness of fly ash in adsorbing copper, nickel and zinc by conducting batch kinetic and isotherm studies. Daniels et al. [115] estimated the achievability of utilizing alkaline fly ash to control acid mine drainage. The results discovered that this technique had potential for industrial scale use.

The chemical composition of bagasse fly ash acquired from sugar industries comprises silica 60.5%, alumina 15.4%, calcium oxide 2.9%, iron oxide 4.9% and magnesium oxide 0.81%. The density, porosity and surface area of this material were 1.01 g/cm^3, 0.36 fractions and 450 m^2/g, respectively. X-ray spectra of bagasse fly ash illustrated the presence of various minerals such as kaolinite, mullite, goethite, α-quartz, γ-alumina and haematite. Scanning electron microscope (SEM) studies indicated flocs and porous nature of the material. An adsorbent dose of 10 g/L was found satisfactory for 100% removal of lead (at $4.8 \times 10^{-4} - 4.83 \times 10^{-3}$ M concentrations of lead) indicating high efficiency of bagasse fly ash [120]. Experiments with actual wastewater from metal finishing plant bearing 17 mg/L Pb(II) provided some successful results. It was found that almost complete removal of Pb(II) from 50 mL of wastewater at pH 3.2 was possible with 4.0 g/L of the adsorbent.

Coal fly ash for the removal of Cu(II) and Cd(II) from wastewaters was used by Papandreou et al. [121]. The data followed Langmuir model with maximum removal after 72 h. Ríos and co-workers [122] used coal fly ash, natural clinker and synthetic zeolites to remove Fe(II), As(III), Pb(II), Zn(II), Cu(II), Ni and Cr(III) metal ions from acid mine drainage of mining industry. According to the authors maximum removals were obtained at pH 7.70–9.43. The order of removal found was Fe(II) > As(III) > Pb(II) > Zn(II) > Cu(II) > Ni(II) > Cr(III). Mohan and Gandhimathi [123] used coal fly ash for the removal of Zn(II), Pb(II), Cd(II), Mn(II) and Cu(II) metal ions from municipal solid waste leachate. The fly ash concentration to achieve maximum removal was found to be 2.0 g/L with the removal efficiencies of 39, 28, 74, 42 and 71% for Cu(II), Mn(II), Pb(II), Zn(II) and Cd(II), respectively. Mishra and Tiwari [124] reported the removal of Cu(II), Co(II) and Ni(II) on sodium hydroxide treated fly ash. The data followed Langmuir equation with the order of removal Cu(II) > Co(II) > Ni(II).

A sewage sludge with a chemical composition similar to that of fly ash to remove NI(II) and Cd(II) from wastewater has been used by Elouear et al. [125]. Composite materials of activated carbon and zeolite from coal fly ash and used them for the removal of Ni(II), Cu(II), Cd(II) and Pb(II) was prepared by Jha et al. [126]. The relative selectivity of metal ions were in the order of Pb(II) > Cu(II) > Cd(II) > Ni(II), with equilibrium uptake capacities of 2.65, 1.72, 1.44 and 1.20 mM/g, respectively. As per authors, Langmuir isotherm fitted well and sorption was thought by ion exchange with Na+. The process was found to be pseudo-second order with rate constants of 0.14, 0.17, 0.21 and 0.20 Lg/mM for the uptake of Pb(II), Cu(II), Cd(II) and Ni(II). The cost of the prepared adsorbent was anticipated to be about US$ 12/ton 20 time lower than the cheapest commercially available carbon costs ~US$ 285/ton. Bagasse fly ash had been redeveloped successfully using acid solutions but this was not an economically worth method. Fly ash achieved from sugar industries is more economic having higher adsorption capacities than the fly ash obtained from thermal power plants. Briefly, sugar industry fly ash has a good potential in the adsorption of metal ions in wastewater. However, the capabilities of this fly ash have not been fully investigated till today and need further extensive research.

2.5.1.14 *Fertilizer Industries Wastes*

The waste slurry generated during the liquid fuel combustion in fertilizer plants can be converted into an inexpensive carbonaceous adsorbent material [127]. Generally, hydrogen peroxide is used for the treatment of these wastes and then heated to 200 °C till the evolution of black soot stopped. The heated product is cooled before activating in presence of air at 450 °C for 1 h yielding an adsorbent of 630 m^2/g surface area. The developed adsorbent has good adsorption capacities for the removal of Cr(VI), Hg(II), Pb(II), Cu(II) and Mo(II) metal ions from metallurgical and electroplating wastewaters. The carbon, alumina and iron oxide 92, 0.4 and 0.6 %, respectively [128] are used to determine the compositions of these adsorbent. X-ray diffraction analysis was used to find out the amorphous nature of the activated carbon. The surface structure of the activated product is observed using a SEM. SEM pictures of the material indicated texture and large porosity of the sample activated in air. The produced activated carbon was found to be quite stable in water, salt solutions, acids and bases. Despite of good future of this waste in the production of low-cost adsorbent, much work has not been carried out and its an extensive field of research under progress.

2.5.2 Preparation of Adsorbents from Inorganic Precursors

The better adsorption capacity, free availability and wide range of applications of some inorganic materials have made them a good low-cost alternative of adsorbents. The selection criteria are similar to that of organic precursors. The study of metal oxides and hydroxides, clays, red mud, zeolites, blast furnace slag, sediment and soil, ore minerals and various other substances is essential in case of inorganic precursors. Therefore, the attempts have been made to describe their cost, performance, adsorption capacities and adsorption kinetics in order to remove the toxic metal ions.

2.5.2.1 Metal Oxides and Hydroxides

The oxides and hydroxides of certain metal ions are consisted of free valences and can appropriately be used as adsorbents to remove the toxic metal ions from water. Therefore, titanium oxide was tested for the removal of Cr(VI) [129] whereas electrolytic manganese dioxide (MnO_2) was found appropriate for the adsorption of Th(I), Pb(II), Cu(II), Bi(III) and Cr(III) at pH 2-5 [130]. The amount of metal ions adsorbed was found to increase with increase in pH for the cases of lead and copper ions. A significant drop in the adsorption of thallium and chromium was also observed between pH 3 and 4. The investigation of silica and alumina to study the adsorption behavior of cadmium and zinc had also been carried out at 25 °C for pH 6-6.5 [131]. To investigate the effect of complexing organic ligands such as citrate, acetate and EDTA on the adsorption of Pb(II), Ni(II) and Cd(II) onto the adsorbent [132], a mixture of dry waste Fe(III)/Cr(III) hydroxide obtained from Southern Petrochemical Industries Corporation Limited (SPIC), Tuticorin, Tamilnadu, India was used as an adsorbent.

Similarly, in another study, the waste of Fe(III)/Cr(III) hydroxide was tested as an adsorbent for the removal of As(III), Hg(III), Cr(III) metal ions by Namasivayam and Senthilkumar [133]. The authors [134] utilized the same industrial waste (Fe(III)/Cr(III) hydroxide) as an adsorbent for the removal of toxic ions and dyes from wastewater and they also studied the same for the adsorption of Ni(II) [135]. It was found that the process was significantly affected by the temperature, pH and initial concentration. The performed desorption studies confirmed that 70% of Ni(II) can be removed from the adsorbent at a pH of 4.0. The different conditions were optimized by Srivastava et al. [136] for the removal of Hg(II) from aqueous solution using the same waste of Fe(III)/Cr(III) hydroxide. The desorption of Hg(II) suggested that it is solubilized in 2% KI to the extent of 65%. The adsorption of two metal ions, i.e. Pb(II) and Cd(II) on hydrous aluminum oxide and hydrous iron oxide indicated that the removal of lead and cadmium is greater on hydrous ferric oxide [Pb = 230 and Cd = 72 mg/g of oxide at pH 4.0] than on hydrous aluminum oxide (Pb = 33 mg and Cd(II) = 31 mg/g of oxide). The investigation of the adsorption of Cd(II) on aluminum oxide in the presence of polyacrylic acid by Floroiu et al. [137] found that the metal oxides and hydroxides are suitable adsorbents in batch process but cannot be used as column materials due to their high cost of production. Moreover, the requirement of the costly and specific operations for the management of the used adsorbent is a serious problem and hence these synthetic wastes cannot be an alternative to commercially activated carbon.

2.5.2.2 Red Mud

Red mud is a waste material produced during the production of alumina when the bauxite ore is subjected to caustic leaching. A large quantity of red mud (1–2 tons red mud per ton of alumina produced) is produced in a typical Bayer process plant and the toxicity and colloidal nature of red mud particles create a severe pollution hazard. There have been many proposals in which red mud is utilized in the manufacture of red mud bricks, as filler in asphalt road construction, as iron ore, and as a source of various minerals. The physical, chemical and mineralogical properties of a red mud have been fully explained with the composition of iron oxide (38.80%), titanium oxide (18.80%), alumina (17.28%), silica (9.64%), and sodium oxide (6.86%) [138]. The density, porosity and surface area were 2.0 g/cm^3, 0.45% fraction, and 108 m^2/g, respectively. The presence of hematite, cancrinite, goethite, rutile, anatase and quartz was confirmed by 'd' spacing values from X-ray diffraction studies. The surface texture and porosity of the material with a texture like aluminum silicates distributed with heavy constituents such as iron [138] was revealed through scanning electron (SE) micrographs of activated red mud.

The feasibility of red mud for wastewater treatment was evaluated by Lopez et al. [139]. Red mud was obtained from Alumina-Aluminio of San Ciprian (Lugo, Spain). A salt-free preparation of red mud had specific surface area of 58 m^2/g. Red mud aggregates were utilized in batch and continuous adsorption experiments for the adsorption of heavy metals *viz.* Ni(II), Cu(II), Cd(II) and Zn(II). These experiments pointed out that, after pretreatment red mud is suitable for the treatment of wastewaters. The removal of toxic metal ions, i.e. copper(II), lead(II) and cadmium(II) from wastewater using red mud as adsorbent was studied by Apak and coworkers [140]. The sorption sequence of metals under consideration was Cu > Pb > Cd. Altundoan et al. [141] used heat treatment and acid treatment methods for red mud to increase its arsenic adsorption capability. Sodalite compounds were leached out due to this treatment. As(III) and As(V) adsorption characteristics of activated red mud were similar to that with raw red mud. Batch adsorption studies had proved that activated red mud in dosages ranging from 20 to 100 g/L could be used successfully to remove arsenic from aqueous solutions. Only few reports are available on the water purification by using red mud due to its recent introduction as a low-cost alternative to activated carbon despite quite good adsorption capacities of red mud.

2.5.2.3 Clays

For many years, fuller's earth has been successfully utilized in many industrial applications all over the world because of its decolorizing properties, low cost and high availability. The negative charge on the surface of the fine grain silicate minerals is responsible for the adsorption capabilities of clay and this negative charge get neutralized by the adsorption of positively charged cations such as heavy metals. Also, the large surface area up to 800 m^2/g has resulted into high adsorption capacity [142] of clay. There are many types of clays but montmorillonite clays are known to have possessed the highest adsorptive capacities as compared to the others. The adsorption process have been tested for many clays and it was observed that the uptake of Hg(II) by montmorillonite is five times greater than kaolinite, [143] wollastonite for nickel, [144] fly ash and wollastonite (1:1 mixture) for Cr(VI) [145]. The effect of a range of commercially available water soluble surfactants on the uptake of Cu(II), Zn(II), Cd(II) and Pb(II) ions by three types of clays *viz.* kaolinite, illite and a montmorillonite was thoroughly examined by Beveridge and Pickering [146]. From the study, it was revealed that the charged surfactants greatly influence the phase distribution process if present at low concentration ($<$0.005% w/v) but they were required at much higher levels of nonionic species ($>$0.1% w/v) to produce the observable effects. The adsorption of lead and cadmium on clay samples (montmorillonite and kaolinite) by Srivastava et al. [147] found that the adsorption of metal ions is greater on montmorillonite and the adsorption capacity increases with increase in pH. The wastewater from a metal finishing plant was treated on clay (montmorillonite) and all the metal ions except Ni(II) were removed with 100% efficiency.

China clay can also be utilized for the removal of As(III) [148] and Zn(II) [149] from aqueous solutions. The adsorption of nickel onto natural clay was studied by the Hawashs and co-workers [150]. Viraraghavan and Kapoor [151] reported the abundance of low-cost bentonite and strongly recommended it as an adsorbent for the removal of heavy metals from wastewaters. Singh et al. [152] investigated the ability of hematite to remove Cr(VI) from aqueous solutions whereas Khan and co-workers [153] studied the adsorption of Cr(III), Cr(VI) and Ag(I) from the aqueous solutions using bentonite. The results of zinc adsorption onto natural bentonite indicated that the adsorption is physical in nature and the operational parameters such agitation speed, solid liquid ratio, temperature, particle size and initial zinc concentration influences the rate of adsorption [154]. In an attempt, to enhance the efficiency of clay for the removal of pollutants from wastewaters, Cadena et al. [142] modified bentonite by replacing the natural exchangeable cations by the organophilic cations, tetramethyl ammonium ion (TMA^+). The prepared adsorbent was examined to adsorb Pb(II) and Cr(VI) from aqueous solutions and the adsorption capacities for the removal of Pb(II) and Cr(VI) by natural bentonite were found to be 6.0 and 5.5 mg/g whereas the one obtained using tailored bentonite were approximately 58 and 57 mg/g for lead and chromium, respectively. The results of the heat and acid treatments on natural bentonite revealed that the adsorption capacity can be improved by heat treatment but the acid treatment decreases the same [155]. The study of the adsorption of Cr(III) from a tanning wastewater on a kaolinite by Tavani and Volzone [156] found that the adsorption of Cr(III) rises steeply at low equilibrium concentrations but results in negligible changes at high equilibrium concentrations. The Cd(II) metal ion was observed to be adsorbed on kaolinite through ion exchange at permanently charged sites on the silanol faces and complexation to aluminol and silanol groups at the crystal edges [157].

2-Mercaptobenzimidazole loaded natural clay can be utilized for the removal of Hg(II) [158]. The adsorption of Hg(II) increased with increasing pH and reached a plateau in pH range 4.0–8.0. The removal of Hg(II) was found to be >99% at an initial concentration of 50 mg/L. Lin and Puls [159] studied the adsorption/desorption and oxidation/reduction of arsenic at clay surfaces. Three types of clay mineral 1:1 layer clays [halloysite (IN), sedimentary M-kaolinite, and weathered EPK-kaolinite]; 2:1 layer clays [illite (MT) and illite/montmorillonite (MT)]; 2>1 layer clay [chlorite (CA)] were studied in this study. The halloysite and the chlorite had much greater As(V) adsorption (25–35 folds) than the other clay minerals.

The removal of Ni(II) industrial effluents on silylated clays by optimizing pH and the desorption process has been studied by Carvalho et al. [160]. The adsorption and desorption of Cd(II) on soils and clay has been compared by Asçi et al. [161]. Authors study reveals that the sorption process is affected by soil pH, initial metal concentration and clay mineralogy. Freundlich model estimated that soils with higher clay content; characterized with smectite as a dominant component; had the greatest sorption capacities and intensities. Cd(II) recovery from the soils was investigated as a function of pH, amount of Cd(II) and rhamnolipid concentration. Tunisian smectite clay was used by Chaari and co-workers [162] for the removal of Pb(II) metal ions. The authors concluded that the adsorption of lead increased with an increase in solution pH from 1.0 to 4.5 and then decreased, slightly between pH 4.5 and 6, and rapidly at pH 6.5 due to the precipitation of Pb(II) ions. The maximum adsorption (25 to 25.44 mg/g) was within 20 min from 25 to 40 °C.

A kaolinite clay modified with 25% (w/w) aluminum sulfate and unmodified kaolin for adsorption of Pb(II) from aqueous solution was used by Jiang et al. [163]. According to the authors, the amount of Pb(II) adsorbed onto modified kaolin (20 mg/g) was more than 4.5 times than that adsorbed onto unmodified kaolin (4.2 mg/g) under the optimized condition. Guerra et al. [164] treated a natural montmorillonite and a synthetic kanemite with 1,4-bis(3-aminopropyl)piperazine reacted with methylacrylate and used them for the removal of Th(IV), Ur(VI) and Eu(III) from aqueous solution. The pH and metal concentration were optimized. The data followed nonlinear adsorption isotherm model. Calcined Bofe bentonite clay for the removal of Ni(II) metal ions was utilized by Vieira et al. [165]. The clay removed nickel with maximum adsorption capacity of 1.91 mg metal/g of clay (20 °C; pH 5.3). U(VI) from aqueous solution using humic acid-immobilized zirconium-pillared clay was removed by Anirudhan et al. [166]. A maximum removal of 97.6% and 94.7% was observed for initial concentrations of 50 and 100 mg/L, respectively at pH 6.0, adsorbent dose of 2.0 g/L and 180 min equilibrium time. The authors accomplished complete removal of U(VI) from 1.0 L of a simulated nuclear industry effluent sample containing 10 mg U(VI) ions using only 1.5 g of adsorbent. From this conversation, it was evident that clays have not been used at large scale due to their limited adsorption capacities. Moreover, it is not easy to prepare clay in granular form which is required in column fillings.

2.5.2.4 Blast Furnace Slag

Steel plants manufacture a hefty volume of granular blast furnace slag as a waste product which causes a disposal problem. It is being used as filler or in the production of slag cement. Recently, it has been transformed into an useful and inexpensive adsorbent for the removal of some toxic pollutants from wastewater. The produced adsorbent demonstrated good adsorption capacities for Cu(II), Ni(II) and Zn(II) ions. The removal of Cu(II), Ni(II), Zn(II), Pb(II) and Cr(III) metal ions using blast furnace slag as adsorbent has been reported by Nehrenheim et al. [167]. Contact time was found most crucial in controlling adsorption among various variables studied. The data followed a pseudo-second-order kinetic model. Blast furnace slag could not be used much for removal of metal ions despite having a great because of its costly pretreatment and poor availability.

2.5.2.5 Zeolites

Zeolites are naturally occurring silicate minerals, which can also be produced at commercial level. Clinoptilolite is probably the most abundant of more than 40 natural zeolite species [168]. This mineral is readily available and inexpensive due to the fact that it occurs in abundance throughout the western United States. Zeolites ion-exchange capabilities are the main cause of their adsorption properties. The three dimensional structure of zeolite possesses large channels containing negatively charged sites. Sodium, calcium, potassium and other positively charged exchangeable ions occupy the channels within the structure and can be replaced by heavy metals. According to Leppert [169], zeolites particularly clinoptilolite exhibits strong affinity for Pb(II) and other heavy metals. The authors revealed that the overall adsorption capacities for zeolites varies for different species but tends to be around 1.5 meq/g (155.4 mg Pb/g zeolite). Clinoptilolite-rich rocks have preferential affinity for Pb(II) over other metal ions. However, zeolites ineffectiveness for Cr(VI) removal has been reported [170].

The different parameters namely zeolite particle size, pH, initial metal ion concentrations, slurry concentration and solution temperature were researched for the sorption of lead and nickel ions from aqueous solution [171]. This study explained that zeolite (phillipsite) tuff is hugely efficient for removing both lead and nickel under all conditions tested. The removal capabilities are higher for lead than for nickel. Natural Bulgarian zeolite was tested for its ability to remove Cu(II) from model wastewater [172]. It was showed that the optimum wastewater to zeolite ratio is 100:1 and the optimum pH value of water to be treated is 5.5 to 7.5. Zeolite modification by treating with NaCl, CH_3COONa and NaOH increased its uptake ability. Copper ions are powerfully unmoved by modified zeolite. The secondary pollution of water caused by its contact with preloaded zeolite is very low [(1.5–2.5% of Cu(II)]. Uptake of Cu(II) by zeolite from wastewater has found to be as effective as Cu(II) by precipitation in its hydroxide form. This fact, together with the correlation found between Cu(II) uptake and the amount of sodium, potassium and calcium released into solution by zeolite showed that the ion exchange sorption played a basic role in Cu(II) uptake by natural zeolite. Langmuir equilibrium constant shows reasonable affinity of zeolite for Cu(II). Values of ΔG° and ΔH° showed the spontaneous and endothermic nature of Cu(II) uptake process by natural zeolite. The iron from aqueous solutions by batch ion exchange with a solid Na-Y zeolite has been removed by Kim et al. [173].

The pH of the iron solutions was low enough to prevent the oxidation of Fe(II) and subsequent hydroxide deposition in the case of Fe(II)/Na-Y exchange system. An increase in the initial Fe(II) concentration, in the range 0.005–0.05 M, lowered the removal efficiency, but the external Fe(II) was favored to the indigenous sodium over the entire concentration range. A maximum Fe(II) recovery of 84% from the maximally exchanged zeolite was achieved using 2.0 M solutions of NaCl while the regenerated Na-Y delivered 68% of the original Fe(II) exchange capacity. Fe(II) recovery was lower from samples exchanged under reflux conditions, while drying the loaded zeolite at 110 °C also suppressed the degree of recovery. Due to the acidity associated with the zeolite/salt slurries, the treatment of Fe(III) solutions with Na-Y was not feasible, which promotes excessive hydroxide deposition and structural disintegration of the zeolite. The natural chabazite zeolite for the removal of the fission products, ^{90}Sr and ^{137}Cs from wastewater and groundwater was utilized by Bostick et al. [174]. The sorbent Ionsiv™ IE-911, a crystalline silicotitanate, is commercially developed and produced by Universal Oil Products (UOP, Des Plaines, Illinois, USA). It was recently tested for fission product removal and found to compare tremendously well against the baseline material.

The removal of Cd(II), Cu(II), Pb(II) and Zn(II) metal ions on Jordanian zeolites from drinking and wastewater samples by optimizing zeolite particle size, ionic strength and initial metal ion concentrations, respectively was performed by Baker et al. [175]. Composite materials of activated carbon and zeolite by activating coal fly ash for removal of Ni(II), Cu(II), Cd(II) and Pb(II) metal ions was prepared by Jha et al. [126]. The relative selectivity of adsorbent was Pb(II) > Cu(II) > Cd(II) > Ni(II), with equilibrium uptake capacities of 2.65, 1.72, 1.44 and 1.20 mM/g, respectively. The overall reaction was pseudo-second order with rate constants of 0.14, 0.17, 0.21 and 0.20 L g/mM for removal of Pb(II), Cu(II), Cd(II) and Ni(II) metal ions, respectively. Zeolite may be good alternatives for activated carbon but these are in equilibrium with metal ions in the nature and, hence, needs pretreatments, which may be costly. Moreover, the adsorption occurs through ion exchangeable channels. Hence, the regeneration is likely by specific chemicals, which may be costly. Small particle size makes clays bad adsorbents for column operations as they may clog them.

2.5.2.6 Sediment and Soil

Sand, sediment and soil also have high free valences, which are accountable for their good adsorption properties for various pollutants. Therefore, many researchers tried to use these adsorbents for the removal of metal ions from aqueous solutions. Bailey et al. used iron-oxide-coated sand for the removal of Cr(VI) in this direction. Different parameters for the adsorption process are optimized by the author. The adsorption of copper, chromate and arsenate onto iron-oxide-coated sand (IOCS) in single-metal and mixed-metal systems are studied by Khaodhiar and coworkers [176]. Copper and arsenate were powerfully adsorbed or formed inner sphere surface complexes with the IOCS surface while chromate was weakly adsorbed or formed an outer sphere surface complex with the IOCS surface. The triple-layer model (TLM) was used and the model shown that each metal used a different number of adsorption sites. For Cu(II), Cr(VI) and As(IV) metals IOCS systems, the equilibrium constants found from single solute systems were not able to forecast adsorption from multisolute systems. The TLM had not presently accounted for the heterogeneity of oxide surface sites and the formation of ternary complexes and/or solid phases that did not exist in single solute systems. Huang [177] studied the mechanism of sediment motion that affected cadmium adsorption on sediment particles was studied in a turbulence tank in the presence and absence of bed mud. The verification of experimental results was carried out by a mathematical model for heavy metal transport developed for the turbulence tank. The transport transformation of heavy metal pollutants in surface waters following the law of convective diffusive of common tracers and the characteristics of fate and transport of sediment motion was confirmed by the model. Variations of dissolved Cd(II) concentrations and suspended particulate Cd(II) concentrations with time and in the water column were measured and computed. Both the experimental measurements and computed results found that it took about 6 h to reach equilibrium condition for cadmium adsorption by sediment particles.

Sediment of Tafna River in northwest of Algeria for the removal of Zn(II) were used by Dali-Youcef et al. [178]. The optimization of effect of various operating variables, viz. initial concentration, sediment dose and contact time have been carried out. The extent of adsorption increased with increase of concentration and decrease of adsorbent dose within 30 min contact time. It was also found that the content of carbonate in sediment increased adsorption indicating the active support material toward zinc ions. The data followed pseudo-second-order mechanism. The thermodynamics of adsorption pointed out spontaneous, endothermic and randomness nature of adsorption. Sediments collected from heavy traffic areas and a residential area with an infiltration type sewage system in Tokyo, Japan for removal of Cu(II), Cd(II) and Zn(II) metal ions from groundwater in Japan was used by Murakami et al. [179]. Soils have been used to remove the environmental pollutants by adsorption.

The removal of Fe(II), Mn(II) and Cr(III) onto a soil as a function of the reaction time, pH, and metal concentration has been reported by Flogeac et al. [180]. The order of removal was Fe(II) > Cr(III) > Mn(II). Wang et al. [181] used two soils as adsorbents for the removal of Zn(II). Three soils for removal of Cd(II) metal ions by enhancing their adsorption capacities by rhamnolipids as biosurfactants were used by Asci et al. [161]. These materials are potential adsorbents due to the fact that these are freely available, good efficiency and performance. The main drawbacks with these materials are that, sometimes they require costly pretreatment, i.e. removal of adherent impurities or coating by some chemicals.

2.5.2.7 Ore Minerals

Ore minerals were found appropriate for the removal of inorganic pollutants by adsorption. The presence of pyrolusite and goethite in ferruginous manganese ore make them the most important for removal of metal ions. It is also a low-cost adsorbent, which had been studied for the removal of arsenic from groundwater [182]. The presence of bivalent cations, *viz.* Ni(II), Co(II) and Mg(II) improved the adsorption capability of this ferruginous ore. It has been fruitfully used for the removal of arsenic from six real groundwater samples containing arsenic in the range of 0.04–0.18 ppm. Arsenic removal was 100% in all cases. The adsorption of U^{VI} on quartz and clinoptilolite, important mineral phases at the proposed US nuclear waste repository at Yucca Mountain, NV, USA was studied by Prikryl et al. [183]. The experiments were conducted at an initial U^{VI} aqueous concentration of ~2.0×10^{-7} M (0.1 M $NaNO_3$ matrix) and over the pH range ~2.5 to ~9.5. U^{VI} solutions were reacted with either quartz or clinoptilolite only or with mixtures of the two minerals removing more than 90% of U^{VI}. The adsorption of Sb(III) from water on hydroxyapatite [$Ca_5(PO_4)(3-x)(CO_3) \times (OH)(1+x)$ $(x=0.3)$] has been studied by Leyva et al. [184]. It was found that more than 95% of the Sb(III) in solution adsorbed to the solid phase by hydroxyapatite in less than 30 min. The equilibrium distribution of Sb(III) (solid vs liquid phase) was characterized by a Langmuir model with $\gamma_{max} = 6.7 \pm -0.1 \times 10^{-8}$ M $(1.4 \pm -0.2 \times 10^{-4}$ M and $K_{ads} = 1.5 \pm 0.2 \times 10^3$ dm^3/L. The granite was utilized for the removal of lead which displayed a typical cation sorption behavior with fractional uptake increasing with increasing pH [185]. Additionally, lead sorption was essentially ionic strength independent, suggesting strong binding and substantial retardation under these circumstances. Not too many reports are available on the adsorption of metal ions from water by ore minerals. The utility of these adsorbents is not sufficient to be used at commercial level. Some other work of removal of inorganic pollutants by low-cost adsorbent is given in Table 2-2.

Table 2-2 Application of Some Low-Cost Adsorbents for Removing Metal Ions

Sl. No	Metal Ions	Types of Water	Adsorbents	pH	Contact Time (minutes)	Removing Capacities	References
1	Ag(II), Cu(II) and Hg(II)	Lab. synthesized water sample	Expanded perlite	6.5	-	Ag: 8.46 mg/g Cu: 1.95 mg/g Hg: 0.35 mg/g	[186]
2	Cd(II)	Lab. synthesized water sample	Modified pine bark	7.0	90	97%	[187]
3	Co(II)	Lab. synthesized water sample	Oscillatoria angustissima	-	15	76%	[188]
4	As(III)	Drinking water	Activated carbon supported nano zero-valent iron	6.5	-	-	[189]
5	Cd(II)	Lab. synthesized water sample	Sugarcane bagasse, Maize corncob and Jatropha oil cake	6.0	60	99.5%	[190]
6	As(III)	Lab. synthesized water sample	Ulothrix cylindricum	-	-	67.2 mg/g	[191]
7	Cd(II)	Lab. synthesized water sample	Hydroxyapatite	-	-	63–83%	[192]
8	As(III), As(V) and Fe(II)	Ground water samples	Tea fungal	-	30–90	-	[193]
9	Au(III), Pt(II) and Pd(II)	-	Glycine-modified chitosan	2.0	-	120.39–169.98 mg/g	[194]
10	Au(I), Pd(II) and Pt(II)	-	DMA paper gel	-	-	0.9–4.6 M/kg	[195]
11	As(V) and As(III)	Effluent of metallurgical industry	Pretreated spent grains	-	-	As(V): 13.39 mg/g As(III): 4.86 mg/g	[196]
12	Au(I), Ag(II) and Pd(II)	-	Nanometer-size titanium dioxide	-	-	11.82–22.63 mg/g	[197]
13	Cd(II) and Pb(II)	Lab. synthesized water sample	Sugar beet pulp	5.3–5.0	70	70–75%	[198]
14	As(III) and As(V)	Lab. synthesized water sample	Macrofungus (Inonotus hispidus)	As(III): 6 As(V): 2	30	As(III): 51.9 mg/g As(V): 59.6 mg/g	[199]

15	Au (I and III)	Aqueous cyanide media	Polyurethane foams	-	-	Au(I): 11.21 mg/g	[200]
16	Cd(II)	Wastewater	Lentinus edodes	-	-	5.58 mM/g	[201]
17	Cd(II)	Ground water and industrial effluents	Calotropis procera	5.0 and 8.0.	5.0	90%	[202]
18	As(V)	-	Cu(II), Ni(II), Co(II)-doped goethite samples	-	30–60.0	19.55 mg/g	[203]
19	Cd(II)	Lab. synthesized water sample	Various types of dried sludge	-	-	0.38 mM/g	[204]
20	Cd(II), Cu(II), Ni(II) and Zn(II)	Lab. synthesized water sample	Biomass of brown algae	4.3–6.5	-	3.0 g/L	[205]
21	Cd(II)	Lab. synthesized water sample	Protonated macroalga Sargassum muticum	4.5	-	-	[206]
22	Cd(II) and Pb(II)	Lab. synthesized water sample	Ethanol-treated waste baker's yeast biomass	-	-	Cd: 31.75 mg/g Pb: 60.24 mg/g	[207]
23	Cd(II) and Zn(II)	-	Aquatic moss Fontinalis antipyretica	5.0	-	Cd: 28.0 mg/g Zn: 14.7 mg/g	[208]
24	Cd(II) and Pb(II)	Lab. synthesized water sample	Sawdust of Pinus sylvestris	5.5	-	Cd: 96% Pb: 98%	[209]
25	Cd(II) and Cr(III)	Wastewater	Alumina or chitosan	-	-	Cd: 99.2% Cr: 83%	[210]
26	Cd(II), Cu(II) and Zn(II)	Lab. synthesized water sample	Cassava waste	-	20–30	60–80%	[211]
27	Cd(II), Cu(II) and Zn(II)	Contaminated water	Petiolar felt-sheath of palm	-	30	-	[212]
28	Cd(II), Cu(II), Ni(II), Pb(II) and Zn(II)	Lab. synthesized water sample	Spruce sawdust	-	-	65%	[213]
29	Cd(II), Cu, Mn, Pb, and Zn	Lab. synthesized water sample	Moroccan stevensite	1.5–7.0	-	-	[214]
30	Cd(II), Cu(II) and Ni(II)	Lab. synthesized water sample	Lemna minor L.	-	40–60	59-Cd: 83 mg/g	[215]
31	Cd(II)	-	Rhodotorula sp. Y11	-	-	19.38 mg/g	[216]
32	Cu(II)	Lab. synthesized water sample	Tectona grandis L.f. leaves powder	-	-	15.43 mg/g	[217]

(Continued)

Table 2-2 Application of Some Low-Cost Adsorbents for Removing Metal Ions—Cont'd

Sl. No	Metal Ions	Types of Water	Adsorbents	pH	Contact Time (minutes)	Removing Capacities	References
33	Cd(II) and Pb(II)	Lab. synthesized water sample	Pleurotus platypus, *Agaricus bisporus* and *Calocybe indica*	6.0, 5.0	-	-	[218]
34	Cd(II), Ni(II), Zn(II), Cu(II), Cr(III) and Pb(II)	Lab. synthesized water sample	Eichhornia crassipes, Vallisneria spiralis and Pistia stratiotes	6.0	-	98%	[219]
35	Cd(II)	Lab. synthesized water sample	Manganese nodule residue	2.0–6.0	-	19.8 mg/g	[220]
36	Cd(II), Pb(II), and U(III)	Lab. synthesized water sample	Powder of poplar leaves and branches	-	-	Cd: 2.3 mg/g Pb: 1.7 mg/g U: 2.1 mg/g	[221]
37	Cd(II) and Pb(II)	Lab. synthesized water sample	Activated alumina	5.0	-	-	[222]
38	Cd(II)	Lab. synthesized water sample	Ceramium virgatum	-	-	39.7 mg/g	[223]
39	Cd(II)	Lab. synthesized water sample	Plant leaves	4.1	-	77–85%	[224]
40	Cd(II)	Effluents	Bone char	-	-	0.57 mM/g	[225]
41	Cd(II) and Cu(II)	Lab. synthesized water sample	Carbonate hydroxylapatite (eggshell waste)	-	-	Cd: 94% Cu: 93.17%	[226]
42	Cd(II) and Pb(II)	-	Grape bagasse	Cd: 7.0 Pb: 3.0	-	Cd: 0.774 mM/g Pb: 0.428 mM/g	[227]
43	Cd(II) and Ga(II)	-	Metallothionein fusion protein	Cd: 5.2 Ga: 6.5	-	26.6–12.1	[228]
44	Cd(II)	Lab. synthesized water sample	Ion imprinted polymer (IIP)	-	50	59.2%	[229]
45	Cd(II), Cr(II), Cu(II), Fe(III), Ni(II), Pb(II) and Zn(II)	Lab. synthesized water sample	Adsorbent CACMM2 extracted from a cactus	-	-	~83%	[230]
46	Cd(II), Cu(II), Pb(II), Ni(II), and Zn(II)	Lab. synthesized water sample	Immobilized phytic acid	-	-	6.6–7.7 mg/g	[231]

47	Cd(II), Hg(II), Cr(II) and Pb(II)	-	Dithizone-anchored poly microbeads	5.0	-	4.3–16.4 mg/g	[232]
48	Cr(III)	Lab. synthesized water sample	Tannin-immobilized mesoporous silica bead	5.5	-	-	[233]
49	Cd(II) and Pb(II)	-	Lobophora variegata	5.0	90	Cd: 1.71 mM/g Pb 1.79 mM/g	[234]
50	Cd(II), Zn(II) and Pb(II)	Lab. synthesized water sample	Brown marine macro algae		10	75%	[235]
51	Cd(II), Cu(II), Fe(II), Se(II), Pb(II), Cr(III) and (VI), Ni(II), Zn(II), Co(II), As(III)	Lab. synthesized water sample	Gum kondagogu (Cochlospermum gossypium),	5.0	-	-	[236]
52	Cd(II) and Pb(II)	-	Pine cone with Fenton reagent	7	Cd: 90 Pb: 105	Cd: 91% Pb: 89%	[237]
53	Cr(VI)	Lab. synthesized water sample	Grape waste	4.0	-	1.91 mol/kg	[238]
54	Cr(VI)	Industrial wastewater	Banana peel	2.0	-	-	[239]
55	Cr(III)	-	Agro-waste biosorbents	4.0	-	12.97 mg/g	[240]
56	Cr (VI)	-	Trichoderma viride biomass	2.0	90	-	[241]
57	Cr (VI)	Lab. synthesized water sample	Agaricus bisporus	1.0	-	-	[242]
58	Cr(VI)	Lab. synthesized water sample	Helianthus annuus (sunflower waste)	2.0	-	76.5%	[243]
59	Cr(VI)	Lab. synthesized water sample	Carbonaceous	-	-	56.49 mg/g	[244]
60	Cr(VI)	Saline solutions	Rhizopus arrhizus	2.0	72 h	78.0 mg/g	[245]
61	Cr(VI)	-	Nostoc muscorum	3.0	-	22.92 mg/g	[246]
62	Cr(VI)	Lab. synthesized water sample	Red pine sawdust	3.0	-	87.7%	[247]
63	Cr(VI)	Wastewater	Maghemite nanoparticles	2.5	15	-	[248]
64	Cr (VI) and Fe(II)	-	Cajanus cajan	2.0 and 2.5	-	~96.05 mg/g	[249]

(Continued)

Table 2-2 Application of Some Low-Cost Adsorbents for Removing Metal Ions—Cont'd

Sl. No	Metal Ions	Types of Water	Adsorbents	pH	Contact Time (minutes)	Removing Capacities	References
65	Cr(III)	Lab. synthesized water sample	Spirogyra sp.	5.0	15–180	81.02%	[250]
66	Cr(VI)	Electroplating wastewater	Rice husk carbon and activated alumina	2–4	60	75.3%	[251]
67	Cr(III)	-	Eggshells	5.0	60	160 mg/g	[251]
68	Cr (VI)	Lab. synthesized water sample	Biomass of two marine strains of Yarrowia lipolytica	1.0	-	-	[252]
69	Cr(VI)	-	Mucilaginous seeds of Ocimum basilicum	1.5	-	205 mg/g	[253]
70	Cr(VI), Cd(II) and Co(II)	Lab. synthesized water sample	Chryseomonas luteola TEM 05	6.0	-	-	[254]
71	Cr(VI) and Cu(II)	Lab. synthesized water sample	Ethylenediamine modified rice hull	Cr: 2.0 Cu: 5.5		Cr: 0.45 mM/g Cu: 0.06 mM/g	[255]
72	Cr(III) and Cr(VI)	Lab. synthesized sample	White, yellow and red sands from the United Arab Emirates	2.0–5.0	-	-	[256]
73	Cr (VI)	Lab. synthesized water sample	Rhizopus nigricans	2.0	8 h	99.2%	[257]
74	Cr(III), Fe(II), Cu(II), Cd(II) and Pb(II)	Ground water	Hybrid inorganic/organic alumina adsorbents	1.0–7.0	-	-	[258]
75	Cr(III), Ni(II) and Cu(II)	Electroplating effluent	Poly (gamma-glutamic acid) (gamma-PGA) derived from Bacillus subtilis NX-2	3.0–5.0	-	Cr(III): 95.11% Ni: 89.32%	[259]
76	Cr(III), Cu(II), Mn(II) and Zn(II)	Lab. synthesized sample	Pseudomonas aeruginosa AT18	5.46–7.72	-	-	[260]
77	Cr(III), Mn(II), Fe(II), Ni(II), Cu(II), Zn(II), Hg(II), Pb(II), Pd(II), Pt(II), Ag(II) and Au(III)	Lab. synthesized sample	Silica gel matrix inorganic–organic composite material	-	-	1.700 mM/g	[261]

78	Cr(VI), Cu(II) and Cd(II)	Lab. synthesized sample	4-vinyl pyridine/2-hydroxyethylmethacrylate monomer mixture grafted poly(ethylene terephthalate) fiber	3.0	-	Cr(VI): 99%	[262]
79	Co(II)	-	Carboxylate-functionalized polyacrylamide grafted lignocellulosics	6.5–9.0	-	99%	[263]
80	Cr(VI), Pb(II), Hg(II) and Cu(II)	Lab. synthesized sample	Sawdust (Acacia arabica)	6.0	-	Cr: 111.61 mg/g Pb: 52.38 mg/g Hg: 20.62 mg/g Cu: 5.64 mg/g	[264]
81	Co(II) and Zn(II)	Lab. synthesized water sample	Various low-cost adsorbents	1.5–9	-	>94%	[265]
82	Co(II), Cu(II), and Zn(II)	-	Cyanobacterium Spirulina platensis	6.0	1–2	63–77%	[266]
83	Cs(III)	-	Chemically modified biomass of marine algae	4.0	30	-	[267]
84	Cu(II)	Lab. synthesized water sample	Microwave stabilized heavy metal sludge	9.2–9.5	-	18 to 28 mg/g	[268]
85	Co(II)	Lab. synthesized water sample	Clay minerals (hectorite)	-	48 h	-	[269]
86	Cu(II)	Polyamine-functionalized adsorbent	Triethylenetetramine and tetraethylenepentamine	-	-	63.44 mg/g	[270]
87	Cu(II)	Lab. synthesized water sample	Marine algae Gelidium and algal composite material	5.3	-	13 mg/g	[271]
88	Cs(III) and Cu(II)	Lab. synthesized water sample	Aluminum-pillared-layered montmorillonites (PILMs)	3.0–8.0	-	-	[272]
89	Cu(II)	Lab. synthesized water sample	Rhizopus oligosporus	2.0–6.0	1–48 h	79.37 mg/g	[273]
90	Cu(II)	Lab. synthesized water sample	Chemically modified chitosan	5.0	-	43.47 mg/g	[274]
91	Cu(II)	Lab. synthesized water sample	Sewage sludge and pomace ashes	-	-	6.98 mg/g	[275]

(Continued)

Table 2-2 Application of Some Low-Cost Adsorbents for Removing Metal Ions—Cont'd

Sl. No	Metal Ions	Types of Water	Adsorbents	pH	Contact Time (minutes)	Removing Capacities	References
92	Cu(II)	Lab. synthesized water sample	Palm shell-activated carbon with polyethyleneimine	5.0	-	75.86%	[276]
93	Cu(II)	-	Marrubium globosum subsp leaves	5.5	-	45.90%	[277]
94	Cu(II)	Lab. synthesized water sample	Soybean straw	6.0	-	0.64 mM/g	[278]
95	Cu(II)	Lab. synthesized water sample	Spirulina platensis	-	-	67.93 mg/g	[279]
96	Cu(II), Cd(II) and Pb(II)	Acidic solutions	Calcium alginate beads from Laminaria digitata	4.5	-	-	[280]
97	Cu(II) and Co(II)	-	Penicillium brevicompactum	5.0	60	-	[281]
98	Cu(II)	Wastewater	Green alga Cladophora fascicularis	-	30	-	[282]
99	Cu(II)	Wastewater	Poly-gamma-glutamic acid	4.0	-	77.9 mg/g	[283]
100	Cu(II) and Pb(II)	Lab. synthesized water sample	Manganese oxide-coated sand II	-	-	Cu: 7.56 mM/g Pb: 9.22 mM/g	[284]
101	Cu(II) and Pb(II)	Wastewater	Calcined starfish (SF) and iron-coated SF (ICSF)	-	150	8600 mg/kg	[285]
102	Cu(II)	-	Cotton boll	5.0	2–24 h	11.40 mg/g	[286]
103	Cu(II) and Pb(II)	Lab. synthesized water sample	Chaff	-	-	Cu: 1.98 mg/g Pb: 6.72 mg/g	[287]
104	Cu(II)	Lab. synthesized water sample	Cladonia rangiformis hoffm	5.0	-	7.6923 mg/g	[288]
105	Cu(II)	Lab. synthesized water sample	Marine algae Padina sp	5.0	30	0.80 mM/g	[289]
106	Cu(II)	-	Capsicum annuum (red pepper) seeds	-	60	4.47×10^{-4} mol/g	[290]
107	Cu(II) and Cd(II)	Lab. synthesized water sample	Activated carbon derived from Ceiba pentandra hulls	6.0	-	Cu: 20.8 mg/g Cd: 19.5 mg/g	[291]
108	Cu(II)	-	Smectitic clay	-	30	94.6 to 96.0%	[292]

109	Cu(II)	Lab. synthesized water sample	Gum arabic modified magnetic nano-adsorbent	-	2	-	[293]
110	Cu(II) and Pb(II)	Lab. synthesized water sample	Iron-coated sand	-	-	Cu: 0.259 mg/g Pb: 1.211 mg/g	[294]
111	Cu(II)	Lab. synthesized water sample	Dried activated sludge	4.0	-	294 mg/g	[295]
112	Cu(II), Pb(II) and Zn(II)	Urban runoff	Mulch (cypress bark, hardwood bark, and pine bark nugget)	5.0 and 6.0	-	-	[296]
113	Cu(II) and Fe(II)	Lab. synthesized water sample	Sawdust of beech, linden and poplar trees	Cu: 3.5 Fe: 5.0	~20	7–8 mg/g	[297]
114	Cu(II), Zn(II) and Cr(III)	-	Schwertmannite	6.0–7.0	-	Cu: 99.3%, Zn: 99.4% Cr: 87.6%	[298]
115	Cu(II) and Pb(II)	Lab. synthesized water sample	Tartaric acid modified rice husk	-	-	Cu: 29 mg/g Pb: 108 mg/g	[299]
116	Cu(II), Cd(II), Zn(II), Ni(II) and Pb(II)	Lab. synthesized water sample	Activated sludge	4.0–5.0	-	-	[300]
117	Cu(II), Zn(II), Cd(II) and Pb(II)	Lab. synthesized water sample	Waste tea and coffee adsorbents	4.0–10.0	-	-	[301]
117	Cu(II)	Lab. synthesized water sample	Rhizopus oryzae	4.0–6.0	-	-	[302]
118	Cu(II), Cd(II) and Pb(II)	Lab. synthesized water sample	Pyrite and synthetic iron sulphide	3.0–6.0	-	-	[303]
119	Cu(II) and Zn(II)	Lab. synthesized water sample	Humic acid (HA)	6.0	-	-	[304]
120	Cu(II), Pb(II) and Cd(II)	Lab. synthesized water sample	Crosslinked carboxymethyl konjac glucomannan	5.0–6.0	20	~41.7 mg/g	[305]
121	Cu(II) and Pb(II)	Wastewater	Waste beer yeast	-	-	Cu: 0.0228 mM/g Pb: 0.0277 mM/g	[306]
122	Cu(II) and Pb(II)	Lab. synthesized water sample	Expanded perlite	-	-	Cu: 8.62 mg/g Pb: 13.39 mg/g	[307]

(Continued)

Table 2-2 Application of Some Low-Cost Adsorbents for Removing Metal Ions—Cont'd

Sl. No	Metal Ions	Types of Water	Adsorbents	pH	Contact Time (minutes)	Removing Capacities	References
123	Cu(II), Cd(II), and Pb(II)	Lab. synthesized water sample	Slag	-	-	Cu: 0.101 mM/g Cd: 0.058 mM/g Pb: 0.120 mM/g	[308]
124	Cu(II), Hg(II) and Pb(II)	Lab. synthesized water sample	1-Acylthiosemicarba-zide activated carbon	3.0	-	48.56–78.20 Mg/g	[309]
125	Cu(II), Ni(II), and Cr(VI)	Lab. synthesized water sample	Sawdust (Quercus coccifera)	Cu: 4.0 Ni: 8.0 Cr: 3.0	-	Cu: 93% Ni: 82% Cr(VI): 84%	[310]
126	Cu(II)	Lab. synthesized water sample	Cladosporium sp.	5.0	-	28.31 mg/g	[311]
127	Cu(II), Zn(II), Pb(II), Ni(II), Cd(II) and Ce(III)	Lab. synthesized water sample	Natural seaweed waste	5.5	-	0.78–0.71 mM/g	[312]
128	Cu(II)	Lab. synthesized water sample	Azolla rongpong ions	3.5	-	-	[313]
129	Cu(II), Pb(II), Zn(II) and Cd(II)	Lab. synthesized water sample	Tourmaline	-	-	Cu: 78.86 mg/g Pb: 154.08 mg/g Zn: 67.25 mg/g Cd: 66.67 mg/g	[314]
130	Cu(II), Ni and Mn	-	Marine algae (Ulva lactuca)	2.0–8.0	-	Cu: 92% Ni: 80% Mn: 75%	[315]
131	Cu(II), Cr(III), Ni(II) and Pb(II)	Lab. synthesized water sample	Meranti sawdust	6.0	-	-	[316]
132	Cu(II), Co(II) and Ni(II)	Lab. synthesized water sample	Chitosan/clinoptilolite composite	-	-	4.209–11.32 Mm/g	[317]
133	Cu(II) and Pb(II)	Lab. synthesized water sample	Dolomite powder	-	-	Cu(II): 8.26 mg Pb(II): 21.74 mg	[318]
134	Cu(II) and Cd(II)	Lab. synthesized water sample	Trichosporon cutaneum strain R57	24 h	-	65%	[319]

135	Cu(II), Zn(II), Pb(II) and Mn(II)	Urban storm water	Sargassum biomass	6.0	50	20.2–214 mg/g	[320]
136	Cu(II) and Cr(VI)	Lab. synthesized water sample	Nano-adsorbent	-	-	Cu: 12.43 mg/g Cr: 11.24 mg/g	[321]
137	Fe(III) and Fe(II)	-	Trimesic acid coated alumina	1.5	-	Fe(III): 26.6 mg/g Fe(II): 8.4 mg/g	[322]
138	Fe(II), Fe(III), and Mn(II), Zn(II)	Lab. synthesized water sample	Lignite	-	-	11.90–34.22	[323]
139	Fe(II), Zn(II), and Cu(II)	Acid mine drainage	Rice husk and Desulfotomaculum nigrificans	-	-	Fe(III): 99% Fe(II): 98% Zn and Cu: 95%	[324]
140	Hg(II)	Lab. synthesized water sample	Nonviable Bacillus sp.	4.5 and 6.0	20	-	[325]
141	Hg(II)	Lab. synthesized water sample	Aspergillus versicolor	5.0–6.0		75.6 mg/g	[326]
142	Hg(II), Cd(II) and Pb(II)	-	Microalgae Chlamydomonas reinhardtii	Hg: 6.0 Cd, Pb: 5.0	60	Hg(II): 72.2 mg/g Cd(II): 42.6 mg/g Pb(II): 96.3 mg/g	[327]
143	Hg(II)	-	Pseudomonas fluorescens BM07	7.0	-	82.25 mg/g	[328]
144	Hg(II)	-	Aminated chitosan bead	-	100	90%	[329]
145	Ni(II)	Lab. synthesized water sample	Dried vegetative cell and spore-crystal mixture of Bacillus thuringiensis var. thuringiensis	-	-	15.7%	[330]
146	Ni(II)	Lab. synthesized water sample	Azadirachta indica leaf powder	5.0	-	92.6%	[331]
147	Ni(II)	Lab. synthesized water sample	Sawdust	2.0–5.0	1.0 h	-	[332]

(*Continued*)

Table 2-2 Application of Some Low-Cost Adsorbents for Removing Metal Ions—Cont'd

Sl. No	Metal Ions	Types of Water	Adsorbents	pH	Contact Time (minutes)	Removing Capacities	References
148	Mn(II), Fe(II), Co(II), Ni(II), Zn(II), Cd(II), Hg(II), Pd(II), Cu(II), Ag(II) and Au(III)	Lab. synthesized water sample	Novel silica gel matrix composite adsorbent	-	-	Cu: 0.472 mM/g Ag: 0.822 mM/g Au: 0.810 mM/g	[333]
149	Ni(II)	-	Baker's yeast	6.75	-	9.8 mg/g	[334]
150	Ni(II)	-	Pine tree materials	8.0	-	80%–97%	[335]
151	Ni(II) and Cu(II)	Lab. synthesized water sample	Pseudevernia furfuracea (L.) Zopf	Ni: 4.0 Cu: 5.0–6.0	-	Ni: 49.87 and Cu: 60.83 mg/g	[336]
152	Ni(II)	Lab. synthesized water sample	Thuja orientalis	-	3 .0	90%	[337]
153	Ni(II) and Zn(II)	Lab. synthesized water sample	Activated carbon from Hevea brasiliensis	-	-	22.03 mg/g	[338]
154	Ni(II) and Cu(II)	Lab. synthesized water sample	Iron oxide-coated sand	2.0–9.0	-	-	[339]
155	Ni(II)	Industrial wastewater	Genetically engineered Escherichia coli JM109	4.0–10.0	10	95%	[340]
156	Ni(II)	Plating Ind. effluents	Activated alumina	9.0	-	-	[341]
157	Ni(II) and Co(II)	Lab. synthesized water sample	Octadecyl silica gel	-	-	96%	[342]
158	Ni(II) and Cr(III and VI)	Lab. synthesized water sample	Cochlospermum gossypium	Ni: 5.0 Cr: 2.0	-	-	[343]
159	Ni(II), Co(II), Cd(II), Cu(II), Pb(II), Cr(III) and (VI)	Lab. synthesized water sample	Activated carbon prepared from apricot stone	1.0–6.0	-	-	[344]
160	Ni(II), Cd(II) and Cr(VI)	Industrial effluents	Metal tolerant Aspergillus niger and Penicillium sp.	-	18 h	Cr: 19.3 mg/g Cd: 19.4 mg/g Ni: 25.05 mg/g	[345]

161	Ni(II)	Lab. synthesized water sample	Lewatit cation-exchange resin	2.0–8.0	30–180	171 mg/g	[346]
162	Ni(II) and Cu(II)	Lab. synthesized water sample	Undaria pinnatifida	3.0–7.0	-	Ni: 24.71 mg/g Cu: 38.82 mg/g	[347]
163	Pb(II)	Lab. synthesized water sample	Seed powder of Prosopis juliflora DC	6.0	-	40.322 mg/g	[348]
164	Pb(II)	Lab. synthesized water sample	Seed husk of Calophyllum inophyllum	4.0	-	-	[349]
165	Pb(II)	Lab. synthesized water sample	Tea wastes	5.0	-	-	[350]
166	Pb(II)	Lab. synthesized water sample	Chitosan, Chitosan-GLA and Chitosan-alginate beads	-	-	34.98, 14.24 and 60.27 mg/g	[351]
167	Pb(II)	Lab. synthesized water sample	Shells of hazelnut and almond	2–9	10–240	-	[352]
168	Pb(II)	Lab. synthesized water sample	Macroreticular poly(vinyl alcohol) (MR-PVA) beads	-	-	213.98 mg/g	[353]
169	Pb(II)	Lab. synthesized water sample	Raw and activated charcoals of Melocanna baccifera Roxburgh (bamboo)	2.0–6.0	15–360	-	[354]
170	Pb(II)	Lab. synthesized water sample	Seaweed Sargassum horneri	-	30	696 mM/g	[355]
171	Pb(II)	Lab. synthesized water sample	Waste biomass of aerial roots of Rhizophora mangle (red mangrove)	5.0	-	-	[356]
172	Pb(II)	Lab. synthesized water sample	Wheat bran	-	-	69.0–87.0 mg/g	[357]
173	Pb(II)	-	Dried waste beer yeast	-	90	94%	[358]
174	Pb(II)	Lab. synthesized water sample	Hydroxyapatites	-	-	330–450 mg/g	[359]
175	Pb(II) and Cd(II)	Tap water and river water	Waste Chinese herb Pang Da Hai	-	-	Pb: 27.1 mg/g Cd: 17.5 mg/g	[360]
176	Pd(II)	Lab. synthesized water sample	Aspergillus niger	2.8–7.2	-	96.21%	[361]
177	Pb(II)	-	Quebracho tannin resin	5.0	-	86.207 mg/g	[355]

(Continued)

Table 2-2 Application of Some Low-Cost Adsorbents for Removing Metal Ions—Cont'd

Sl. No	Metal Ions	Types of Water	Adsorbents	pH	Contact Time (minutes)	Removing Capacities	References
178	Pb(II)	Lab. synthesized water sample	Chemically modified carbon adsorbents	-	-	79.43–98.89%	[362]
179	Pb(II) and Cu(II)	-	Penicillium simplicissimum immobilized with loofa sponge	5.0	60	152.6 and 112.3 mg/g	[363]
180	Pb(II), Cu(II) and Cd(II)	Wastewater	Phomopsis sp.	4–6	24 h	110–870 mM/g	[364]
181	Pb(II) and Cd(II)	Lab. synthesized water sample	Rhizobacteria, Azotobacter chroococcum and Bacillus megaterium	-	-	Pb: 59.8% Cd: 75.6%	[365]
182	Pb(II), Cd(II) and Ni(II)	Lab. synthesized water sample	Sawdust of walnut	-	60	-	[366]
183	Pb(II) and Cu(II)	-	EDTAD-modified baker's yeast	2.7 to 6.0	-	Pb: 192.3 mg/g Cu: 65.0 mg/g	[367]
184	Pb(II) and Cd(II)	Aqueous solution	Microparticles of dry plants	-	-	90%	[368]
185	Pb(II), Ag(II), Cs(III) and Sr(II)	Lab. synthesized water sample	Brewery's waste biomass	-	-	0.076–0.413 mM/g	[369]
186	Pb(II) and Cd(II)	Aqueous solutions	Grape stalk waste	5.5	-	Pb(II): 0.241 mM/g Cd(II): 0.248 mM/g	[370]
187	Pb(II), Cd(II), Cu(II) and Ni(II)	Lab. synthesized water sample	Anaerobic granular biomass	-	30	60–255 mg/g	[371]
188	Pb(II), Cd(II) and Cu(II)	Lab. synthesized water sample	Ulmus carpinifolia and Fraxinus excelsior	2.0–5.0	-	69.5–201.1 mg/g	[372]
189	Pb(II), Cu(II), Cd(II), Zn(II) and Ni(II)	Lab. synthesized water sample	Tobacco dust	<2.0	-	24.5–39.6 mg/g	[373]

190	Pb(II), Ni(II), Cd(II) and Zn(II)	Lab. synthesized water sample	Fungus Mucor rouxii	6.0	-	20.49–53.75	[374]
191	Pb(II), Cd(II), Fe(II), La(III) and Ce(III)	-	Laminaria japonica with epichlorohydrin	-	-	0.87–1.35 M/kg	[375]
192	Pb(II), Cd(II), Ni(II) and Zn(II)	Lab. synthesized water sample	Azolla filiculoides	10.5	-	-	[376]
193	Pb(II), Zn(II), Cu(II) and Cd(II)	Lab. synthesized water sample	Activated carbon of Phaseolus aureus hulls	7.0–8.0	-	15.7–6.021.8 mg/g	[377]
194	Pb(II), Cu(II) and Cd(II)	Industry and wastewater	Enterobacter sp.	3.0	-	32.5–50 mg/g	[378]
195	Pb(II)	Mtal solutions	Crop milling waste	-	30	19.56–49.97 mg/g	[379]
196	Pb(II) and Cr(III)	Lab. synthesized water sample	Lichen (Parmelina tiliaceae)	-	-	Pb: 75.8 mg/g Cr: 52.1 mg/g	[380]
197	Pb(II), Cd(II), and Cu(II)	Lab. synthesized water sample	Chitosan	4.0		0.010–0.036 mM/g	[381]
198	Pb(II), Cd(II), Zn(II) and Cu(II)	Lab. synthesized water sample	Rice husk activated (RHA)	1.0–10.0	-	-	[382]
199	Pb(II) and Cu(II)	-	Exopolysaccharide	4.5–6.0	-	Pb: 243.3 mg/g Cu: 36.7 mg/g	[383]
200	Pb(II), Cu(II) and Zn(II)	Lab. synthesized water sample	Polysaccharide produced from Bacillus firmus	-	10	Pb: 98.3% Cu: 74.9% Zn: 61.8%	[384]
201	Pt(II) and Pd(II)	Lab. synthesized water sample	Thiourea-modified Chitosan microspheres	2.0	-	129.9 mg/g	[385]
202	Sr(II)	Lab. synthesized water sample	Bone char	4.0–10.0	-	0.271 mM/g	[386]
203	Sn(II)	Lab. synthesized water sample	Solid-phase Quercetin	1.8–2.2	-	-	[387]
204	U(III)	Wastewater	Citrobacter freundii	-	30	-	[388]

(*Continued*)

Table 2-2 Application of Some Low-Cost Adsorbents for Removing Metal Ions—Cont'd

Sl. No	Metal Ions	Types of Water	Adsorbents	pH	Contact Time (minutes)	Removing Capacities	References
205	Pt(II), Pd(II) and Au(II)	Lab. synthesized water sample	L-lysine modified crosslinked chitosan resin	1.0–2.0	-	70.34–129.26 mg/g	[389]
206	V(II), Re(II), Mo(III), Ge(II), As(V), Cd(II), Hg(II), Al(III), Pb(II), Fe(II) and Cu(II)	Copper mine acidic residual waters	Modified Pinus radiata bark and tannins	3.0	-	7.0–83%	[390]
207	U(III)	-	Cystoseria indica algae	4.0	-	233 mg/g	[383]
208	Zn(II), Pb(II) and Co(II)	Lab. synthesized water sample	Natural Jordanian sorbent	4.5–7.0	-	0.320–2.860 mM/g	[391]
209	Zn(II)	Polluted water	PVA/EDTA resin	6.0	30	-	[392]
210	Zn(II) and Cd(II)	Lab. synthesized water sample	Young brown coals YBC (lignite), humic acids (HAs) and commercial humic acid	5.0–5.7	-	-	[393]
211	Zn(II)	-	Botrytis cinerea	5.0–6.0	-	12.98 mg/g	[394]
215	Zn(II) and Cu(II)	Industrial wastewater	Dead biosludges	-	-	Zn: 92.61% Cu: 83.77%	[395]
213	Zn(II), Cu(II) and Co(II)	Lab. synthesized water sample	Oscillatoria angustissima	4.0–5.0	-	0.12–33 mM/g	[396]
214	Zn(II)	Industrial wastes	Natural zeolites	4.0–6.0	-	-	[397]
215	Zn(II)	Lab. synthesized water sample	Azadirachta indica bark	6.0	-	33.49 mg/g	[398]
216	Zn(II)	Lab. synthesized water sample	Synthesized hydroxyapatites	6.0–8.0	20	98%	[399]
217	Zn(II)	-	Saccharomyces cerevisiae	5.65	0–38 h	74.8–654.8 mM/g	[400]
218	Zn(II)	Eelectroplating wastewater	Wood sawdust and sugarcane bagasse modified ith EDTA dianhydride	-	-	105 mg/g	[401]
219	Zn(II) and Cu(II)	Lab. synthesized water sample	Sugar beet pulp and fly ash	4.0–6.0	60	4.0–6.0	[402]

Conclusions

It is impracticable today to point out a particular adsorbent that will be capable to substitute the ubiquitous activated carbon adsorbent. The individual sections discussed the critical evaluation and comparison of all the possible replacement adsorbents. Briefly, adsorption has been used comprehensively for the uptake of toxic metal ions from ground, surface and wastewaters via batch experiments. Many workers tried to develop economical substitutes to activated carbon. After the analytical observation and comparison of the adsorbents; based on the performance, adsorption capabilities and cost; it was revealed that the most significant and economically feasible adsorbents are fly ashes, red mud and sand because of their free accessibility and good adsorption capacities. The cost of these adsorbents only involves the transportation and simple pretreatment. Moreover, the management of the exhausted adsorbents is the serious and testing problem for the scientists and it is very fascinating to note that heavy metal loaded fly ashes, red mud and sand can still be used as building materials, which make these adsorbents ideals. The adsorption of metal ion species is very vital for the toxicologists and, hence, such types of economical adsorbents should be developed more. Moreover, the usage variety of the adsorbent should be wide with a huge selectivity so that a single adsorbent could be used for greater number of toxic metal ions. Most frequently, the removal capacities of these potential replacement adsorbents have been investigated by using batch operations and, hence, there is a enormous need to transfer batch conditions to column operations so that the preferred columns can be used for water purification and recycling purposes at big scale. To make the adsorption technique more acceptable and applicable, for the removal of toxic metal ions, the enlargement of more useful columns and their applications are required. Basically, column operations are the demand of today and, hence, these should be carried out first at pilot scale plant followed by large community level. In addition, more economical adsorbents should be invented and utilized in packing the columns for the removal of a wide range of toxic metal ions. All the column experiments should be designed by keeping in mind the cost, performance, energy, kinetics and regeneration aspects.

References

[1] L.B. Franklin, Wastewater Engineering: Treatment, Disposal and Reuse, McGraw Hill Inc., New York, 1991.

[2] D.Z. John, Handbook of Drinking Water Quality: Standards and Controls, Van Nostrand Reinhold, New York, 1990.

[3] Toxic Substance Control Act, US EPA, 1984, III, 344.

[4] D. Brusick, Genotoxicity of phenolic antioxidants, Toxicol. Indust. Health 9 (1993) 223–230.

[5] J.S. Mattson, H.B. Mark, Activated Carbon Surface Chemistry and Adsorption from Aqueous Solution, Marcel Dekker Inc., New York, 1971.

[6] D. Reichenberg, Properties of ion exchange resins in relation to their structure. III. Kinetics of exchange, J. Am. Chem. Soc. 75 (1953) 589–594.

[7] S.J.T. Pollard, G.D. Fowler, C.J. Sollars, R. Perry, Low-cost adsorbents for waste and wastewater treatment: a review, Sci. Total Environ. 116 (1992) 31–52.

[8] M.J. Holden, Manufacture and uses of activated carbon, J. Effl. Water Treat. 22 (1982) 27–35.

[9] S.D. Faust, O.M. Aly, Chemistry of Water Treatment, Stoneham, Butterworth, 1983.

[10] P.N. Cheremisinoff, F. Ellerbush, Carbon Adsorption Hand Book, Ann Arbor Science Pub. Michigan, 1979.

[11] W.J. Weber Jr., B.M. Vanvliet, Fundamental Concepts for Applications of Activated Carbon in Water and Wastewater Treatment, in Activated Carbon Adsorption of Organics from the Aqueous Phase, in: I.H. Suffet, M.J. McGuire (Eds.), vol. 16, Ann Arbor Science, Michigan, 1980.

[12] W.J. Lyman, Applicability of Carbon Adsorption to the Treatment of Hazardous Industrial Waste, in Carbon Adsorption Handbook, in: P.N. Cheremisinoff, F. Ellerbusch (Eds.), Ann Arbor Science Publisher, Michigan, 1987.

[13] V.K. Gupta, I. Ali, Adsorbents for Water Treatment, Low-Cost Alternatives to Carbon, Encyclopedia of Surface and Colloid Science, in: H. Arthur (Ed.), vol. 1, Marcel Dekker Inc., New York, 2002, p. 136.

[14] R.P. Bansal, J.P. Donnet, F. Stoeckli, Active Carbon, Marcel Dekker Inc., New York, 1988.

[15] M. Streat, J.W. Patrick, M.J. Camporro-Perez, Sorption of phenol and p-chlorophenol from water using conventional and noval activated carbons, Water Res. 29 (1995) 467–472.

[16] F. Rodriguez-Reinoso, Preparation and characterization of activated carbons, in: Proceeding of the NATO Advanced Study Institute on Carbon and Coal Gasification, NATO ASI Series E, Martinus Nijhoff Publisher, Netherlands, 1986, p. 105.

[17] T. Wigmans, Industrial aspects of production and use of activated carbons, Carbon 27 (1989) 13–22.

[18] E. Fitzer, K. Mueller, W. Schaefer, The chemistry of the pyrolytic conversion of organic compounds to carbon, in: P.L. Walker (Ed.), The Chemistry and Physics of Carbon, vol. 237, Marcel Dekker Inc., New York, 1971.

[19] D.J. Hucknall, Coke and Carbon Formation in Chemistry of Hydrocarbon Combustion, Chapman and Hall, London, 1985.

[20] I.C. Lewis, Chemistry of carbonization, Carbon 20 (1982) 519–527.

[21] M. Smisek, S. Cerny, Active Carbon Manufacture Properties and Applications, Elsevier, London, 1970.

[22] O. Kadlec, A. Varhanikova, A. Zukal, Structure of pores of active carbons prepared by water vapour and zinc dichloride activation, Carbon 8 (1970) 321–331.

[23] H. Marsh, B. Rand, The process of activation of carbons by gasification with CO_2-I. Gasification of pure polyfurfuryl alcohol carbon, Carbon 9 (1971) 47–61.

[24] P.L. Walker, Carbon gasification – a ubiquitous reaction of great importance of mankind, in: Proceedings of the NATO Advanced Study Institute on Carbon and Coal Gasification, NATO ASI Series E, vol. 105, Martinus Nijhoff Publisher, Netherlands, 1986.

[25] ENDS, Government get conflicting advice on options to reduce waste tyre mountains, ENDS Report 186 (1990) 9.

[26] A. Lucchesi, G. Maschio, Semi active carbon and aromatics produced from the pyrolysis of scrap tyres, Conserv. Recyc. 6 (1983) 85–90.

[27] S. Ogassawara, S.M. Kuroda, N. Wakao, Preparation of activated carbon by thermal decomposition of used automotive tyres, Ind. Eng. Chem. Res. 26 (1983) 2552–2256.

[28] P.S. Bryant, J.N. Petersen, J.M. Lee, T.M. Brouns, Sorption of heavy metals by untreated red fir sawdust, Appl. Biochem. Biotechnol. 34 (1992) 777–788.

[29] J.E. Sabadell, R.J. Krack, Adsorption of heavy metals from wastewater and sludge on forest residuals and forest produce waste, water's interface with energy, Air and Solids, in: Proceedings of the Second National Conference on Complete Water Reuse, 4–8 May, Chicago, IL, 1975.

[30] R.C. Vaishya, S.C. Prasad, Adsorption of copper (II) on sawdust, Ind. J. Environ. Protect. 11 (1991) 284–289.

[31] M.A. Zarraa, A study on the removal of chromium (VI) from waste solutions by adsorption on to sawdust in stirred vessels, Adsorp, Sci. Technol. 12 (1995) 129–138.

[32] M. Ajmal, R. Ali-Khan-Rao, B. Ara-Siddiqui, Studies on removal and recovery of Cr (VI) from electroplating wastes, Water Res. 30 (1996) 1478–1482.

[33] C. Raji, G.N. Manju, T.S. Anirudhan, Removal of heavy metal ions from water using sawdust-based activated carbon, Ind. J. Eng. Mater. Sci. 4 (1997) 254–260.

[34] S.K. Srivastava, A.K. Singh, A. Sharma, Studies on the uptake of lead and zinc by lignin obtained from black liquor – a paper industry waste material, Environ. Technol. 15 (1994) 353–361.

[35] M.S. Masri, F.W. Reuter, M. Friedman, Binding of metal cations by natural substances, J. Appl. Polymer Sci. 18 (1974) 675–681.

[36] C.M. Flynn Jr., T.G. Carnahan, R.E. Lindstrom, Adsorption of heavy metal ions by xanthated sawdust, Report of Investigation # 8427. United States Bureau of Mines, 1980.

[37] R.M. Bricka, O.D. Hill, Metal immobilization by solidification of hydroxide and xanthate sludges, in: P.L. Cote, T.M. Gilliam (Eds.), Environmental Aspects of Stabilization of Hazardous and Radioactive Wastes. ASTM STP 1033, American society for testing and materials, Philadelphia, 1989.

[38] V. Tare, S. Chaudhari, M. Jawed, Comparative evaluation of soluble and insoluble xanthate process for heavy metal removal from wastewaters, Water Sci. Technol. 26 (1992) 237–246.

[39] M. Jawed, V. Tare, Application of starch xanthates for cadmium removal: a comparative evaluation, J. Appl. Polymer Sci. 42 (1991) 317–324.

[40] R.E. Wing, W.E. Rayford, Heavy metal removal processes for plating rinse waters, Proceedings of the 32nd Industrial Waste Conference May 10–12, Purdue University, Lafayette, 1977.

[41] R.E. Wing, Dissolved heavy metal removal by insoluble starch xanthate, Environ. Prog. 2 (1983) 269–272.

[42] M.J. Randall, R.L. Bermann, V. Garrett, A.C. Waiss Jr., Use of bark to remove heavy metal ions from waste solutions, J. Forest Prod. 24 (1974) 80–84.

[43] G. Vazquez, G. Antorrena, J. Gonzalez, M.D. Doval, Adsorption of heavy metal ions by chemically modified '*Pinus pinaster*' bark, Bioresource Technol. 48 (1994) 251–255.

[44] M.M. Alves, C.G. Gonzalez Beca, R. Guedes de Carvalho, J.M. Castanheria, M.C. Sol Pereira, L.A.T. Vasconcelos, Chromium removal in tannery wastewaters 'polishing' by *Pinus sylvestris* bark, Water Res. 27 (1993) 1333–1338.

[45] J.M. Randall, E. Hautala, A.C. Waiss Jr., Removal and recycling of heavy metal ions from mining and industrial waste streams with agricultural by-products, in: Proceedings of the Fourth Mineral Wastes Utilization Symposium, 7–8 May, Chicago, IL, 1974.

[46] J.M. Randall, E. Hautala, G. McDonald, Binding of heavy metal ions by formaldehyde polymerised peanut skins, J. Appl. Polymer Sci. 22 (1978) 379–389.

[47] Y. Orhan, H. Buyukgungor, The removal of heavy metals by using agricultural wastes, Water Sci. Technol. 28 (1993) 247–255.

[48] R.U. Edgehill, G.Q. Lu, Adsorption characteristics of carbonized bark for phenol and pentachlorophenol, J. Chem. Technol. Biotechnol. 71 (1998) 27–34.

[49] S. Al-Asheh, Z.J. Duvanjak, Sorption of cadmium and other heavy metals by pine bark, Hazard. Mater. 56 (1997) 35–51.

[50] R.A. Guedes de Carvalho, C.G. Gonzalez-Beca, M.N.O. Sampaio, M.C. Neves, A. Sol Pereira, A. Macedo, Use of pine bark for preparation of activated carbon and as a soil conditioner, Agric. Wastes. 9 (1984) 231–238.

[51] M. Aoyama, M. Tsuda, Adsorption of trivalent chromium from dilute solution by conifer leaves, Wood Sci. Technol. 34 (2000) 55–63.

[52] S.J. Patil, A.G. Bhole, G.S. Natarajan, Scavenging of Ni(II) metal ions by adsorption on PAC and babhul bark, J. Environ. Sci. Eng. 48 (2006) 203–208.

[53] Y. Kalmykova, A.M. Strömvall, B.M. Steenari, Alternative materials for adsorption of heavy metals and petroleum hydrocarbons from contaminated leachates, Environ. Technol. 29 (2008) 111–122.

[54] M.V. Subbaiah, Y. Vijaya, N.S. Kumar, A.S. Reddy, A. Krishnaiah, Biosorption of nickel from aqueous solutions by Acacia leucocephala bark: kinetics and equilibrium studies, Colloids Surf. Biointerfaces 74 (2009) 260–265.

[55] O. Khokhotva, S. Waara, The influence of dissolved organic carbon on sorption of heavy metals on urea-treated pine bark, J. Hazard. Mater. 173 (2010) 689–696.

[56] A.K. Bhattacharya, C. Venkobachar, Removal of Cd(II) by low cost adsorbents, J. Environ. Eng. 110 (1984) 110–122.

[57] K. Srinivasan, N. Balasubramanian, T.V. Ramakrishna, Studies on chromium removal by rice husk carbon, Ind. J. Environ. Health 30 (1998) 376–387.

[58] D. Roy, P.N. Greenlaw, B.S. Shane, Adsorption of heavy metals by green algae and ground rice hulls, J. Environ. Sci. Health A 28 (1993) 37–50.

[59] M.R. Mostafa, Adsorption of mercury, lead and cadmium ions on modified activated carbon, Adsorp. Sci. Technol. 15 (1997) 551–557.

[60] X. Guo, S. Zhang, X.Q. Shan, Adsorption of metal ions on lignin, J. Hazard. Mater. 151 (2008) 134–142.

[61] Y. Wu, S. Zhang, X. Guo, H. Huang, Adsorption of chromium(III) on lignin, Bioresour. Technol. 99 (2008) 7709–7715.

[62] B. Pejic, M. Vukcevic, M. Kostic, P. Skundric, Biosorption of heavy metal ions from aqueous solutions by short hemp fibers: effect of chemical composition, J. Hazard. Mater. 164 (2009) 146–153.

[63] Y.S. Ho, D.A.J. Wase, C.F. Forster, The adsorption of divalent copper ions from aqueous solution by sphagnum moss peat, Trans. I. Chem. 72B (1994) 185–194.

[64] Mineral Resources Institute Technical Report Series, Use of Alabama peat as an adsorbent for heavy metals, MRI Technical Report Series, T.R. No. 12, 1985.

[65] J.K. McLelland, C.A. Rock, Pretreating landfill leachate with peat to remove metals, Water, Air Soil Pollut. 37 (1988) 203–215.

[66] X.H. Chen, T. Gosset, Batch copper binding and exchange properties of peat, Water Res. 24 (1999) 1463–1471.

[67] D.C. Sharma, C.F. Forster, Removal of hexavalent chromium using sphagnum moss peat, Water Res. 27 (1993) 1201–1208.

[68] S.V. Kertman, G.M. Kertman, Z.H.S. Chibrikova, Peat as a heavy metal sorbent, J. Appl. Chem. 66 (1993) 465–466.

[69] T. Viraraghavan, G.A.K. Rao, Adsorption of cadmium and chromium from wastewater by peat, Inter J. Environ. Studies, Sec. A 44 (1993) 9–27.

[70] Y.S. Ho, G. McKay, Kinetics model for lead (II) sorption onto peat, Adsorp. Sci. Technol. 16 (1998) 243–255.

[71] Y. Kalmykova, A.M. Strömvall, B.M. Steenari, Adsorption of Cd(II), Cu(II), Ni(II), Pb(II) and Zn(II) on Sphagnum peat from solutions with low metal concentrations, J. Hazard. Mater. 152 (2008) 885–891.

[72] S.H. Gharaibeh, Y.W. Abu-El-Shar, M.M. Al-Kofahi, MM Removal of selected heavy metals from aqueous solutions using processed solid residue of olive mill products, Water Res. 32 (1998) 498–502.

[73] F.D. Rodriguez-Reinso, J. Lopez-Gonzalez, C. Berenguer, Activated carbons from almond shells – I: preparation and characterization by nitrogen adsorption, Carbon 20 (1982) 513–518.

[74] M.A. Ferro-Garcia, J. Rivera-Utrilla, J. Rodriguez-Gordillo, I. Bautiasta-Toledo, Adsorption of zinc, cadmium and copper on activated carbon obtained from agricultural by-products, Carbon 26 (1989) 363–373.

[75] C. Namasivayam, K. Periasamy, Bicarbonate-treated Peanut Hull. Carbon for mercury(I1) removal from aqueous solution, Water Res. 27 (1993) 1663–1668.

[76] K. Periasamy, C. Namasivayam, Adsorption of Pb (II) by peanut hull carbon from aqueous solution, Sep. Sci. Technol. 30 (1995) 2223–2237.

[77] K. Periasamy, C. Namasivayam, Process development for removal and recovery of cadmium from wastewater by a low-cost adsorbent: adsorption rates and equilibrium studies, Ind. Eng. Chem. Res. 33 (1994) 317–320.

[78] K. Srinivasan, P.B.S. Rao, A. Ramadevi, Studies on characteristics of carbon obtained from tamarind nuts, Ind. J. Environ. Health 30 (1988) 303–305.

[79] J.W. Hassler, Activated Carbon, Leonard Hill, London, 1967.

[80] S.K. Banerjee, S. Majumdar, A.C. Dutta, A.K. Roy, S.C. Banerjee, D.K. Banerjee, Activated carbon from coconut shell, Ind. J. Technol. 14 (1976) 45–49.

[81] G.J. Alaerts, V. Jitaturunt, P. Kelderman, Use of coconut shell-based activated carbon for chromium (VI) removal, Water Sci. Technol. 21 (1989) 1701–1704.

[82] Q. Mortley, W.A. Mellowes, S. Thomas, Activated carbons from materials of varying morphological structure, Thermochim. Acta 129 (1988) 173–186.

[83] J. Laine, A. Calafat, M. Labady, Preparation and characterization of activated carbons from coconut shell impregnated with phosphoric acid, Carbon 27 (1989) 191–195.

[84] G.N. Manju, C. Raji, T.S. Anirudhan, Evaluation of coconut husk carbon for the removal of arsenic from water, Water Res. 32 (1998) 3062–3070.

[85] P.S. Rao, Kinetic studies on adsorption of chromium by coconut shell carbons from synthetic effluents, J. Environ. Sci. Health, Part A Environ. Sci. Eng. 27 (1992) 2227–2241.

[86] G.L. Roorer, T.Y. Hsien, J.D. Way, Synthesis of porous-magnetic chitosan beads for removal of cadmium ions from wastewater, Ind. Eng. Chem. Res. 32 (1993) 2170–2178.

[87] E. Onsoyen, O. Skaugrud, Metal recovery using chitosan, J. Chem. Technol. Biotechnol. 49 (1990) 395–404.

[88] T.C. Yang, R.R. Zall, Adsorption of metals by natural polymers generated from seafood processing wastes, Ind. Eng. Chem. Prod. Res. Dev. 23 (1984) 168–172.

[89] R. Tseng, F. Wu, R. Juang, Pore structure and metal adsorption ability of chitosans prepared from fishery wastes, J. Environ. Sci. Health, Part A, Environ. Sci. Eng. 34 (1999) 1815–1828.

[90] V.M.D. Chui, K.W. Mok, C.Y. Ng, B.P. Luong, K.K. Ma, Removal and recovery of copper (II), chromium (III) and nickel (II) from solutions using crude shrimp chitin packed in small columns, Environ. Int. 22 (1996) 463–468.

[91] C.Y. Kim, H.M. Choi, H.T. Cho, Effect of deacetylation on sorption of dyes and chromium on chitin, J. Appl. Polym. Sci. 63 (1997) 725–736.

[92] A. Baran, E. Biçak, S.H. Baysal, S. Onal, Comparative studies on the adsorption of Cr(VI) ions on to various sorbents, Bioresour. Technol. 98 (2007) 661–665.

[93] R. Jayakumar, M. Rajkumar, H. Freitas, N. Selvamurugan, S.V. Nair, T. Furuike, H. Tamura, Preparation, characterization, bioactive and metal uptake studies of alginate/phosphorylated chitin blend films, Int. J. Biol. Macromol. 44 (2009) 107–111.

[94] K. Vijayaraghavan, Y.S. Yun, Bacterial bio-sorbents and bio-sorption, Biotechnol. Adv. 26 (2008) 266–291.

[95] M.W. Wilson, R.G. Edyvean, Biosorption for the removal of heavy metals from industrial wastewaters, Inst. Chem. Eng. Symp. Ser., Environ. Biotechnol. (1994) 89–91.

[96] Z.R. Holan, B. Volesky, I. Prasetyo, Biosorption of cadmium by biomass of marine algae, Biotechnol. Bioeng. 41 (1993) 819–825.

[97] A. Leusch, Z.R. Holan, B. Volesky, Biosorption of heavy metals [Cd(II), Cu(II), Ni(II), Pb(II), Zn(II)] by chemically reinforced biomass of marine algae, J. Chem. Technol. Biotechnol. 62 (1995) 279–288.

[98] L.K. Jang, D. Nguyen, G.G. Geesey, Effect of pH on the absorption of Cu(II) by alginate gel, Water Res. 29 (1995) 315–321.

[99] M.M. Araujo, J.A. Teixeira, Trivalent chromium sorption on alginate beads, Int. Biodeterior. Biodegrad. 40 (1997) 63–74.

[100] K.H. Chu, M.A. Hashin, S.M. Phang, V.B. Samuel, Biosorption of cadmium by algal biomass: adsorption and desorption characteristics, Water Sci. Technol. 35 (1997) 115–122.

[101] D.J. Malik, V.J. Strelko, M. Streat, A.M. Puziy, Characterization of novel modified active carbons and marine algal biomass for the selective adsorption of lead, Water Res. 36 (2002) 1527–1538.

[102] D. Schmitt, A. Muller, Z. Csogor, F.H. Frimmel, C. Posten, The adsorption kinetics of metal ions onto different microalgae and siliceous earth, Water Res. 35 (2001) 779–785.

[103] C. Hall, D.S. Wales, M.A. Keans, The potential of two strains of *Pseudomonas syringae* (blue and brown) to remove copper from aqueous solutions, Sep. Sci. Technol. 36 (2001) 223–240.

[104] A. Sari, M. Tuzen, Equilibrium, thermodynamic and kinetic studies on aluminum bio-sorption from aqueous solution by brown algae (*Padina pavonica*) biomass, J. Hazard. Mater. 171 (2009) 973–979.

[105] A. Rathinam, B. Maharshi, S.K. Janardhanan, R.R. Jonnalagadda, B.U. Nair, Biosorption of cadmium metal ion from simulated wastewaters using Hypnea valentiae biomass: a kinetic and thermodynamic study, Bioresour. Technol. 101 (2010) 1466–1470.

[106] H. Pahlavanzadeh, A.R. Keshtkar, J. Safdari, Z. Abadi, Biosorption of nickel(II) from aqueous solution by brown algae: equilibrium, dynamic and thermodynamic studies, J. Hazard. Mater. 175 (2010) 304–310.

[107] A. Esmaeili, S. Ghasemi, J. Sohrabipour, Biosorption of copper from wastewater by activated carbon preparation from alga Sargassum species, Nat. Prod. Res. 24 (2010) 341–348.

[108] H. Niu, X.S. Xu, J.H. Wang, Removal of lead from aqueous solutions by penicillium biomass, Biotechnol. Bioeng. 42 (1993) 785–787.

[109] K.A. Jn Matis, A.I. Zouboulis, Flotation of cadmium loaded biomass, Biotechnol. Bioeng. 44 (1994) 354–360.

[110] U. Yetis, A. Dolek, F.B. Dilek, G. Ozcengiz, The removal of lead by *Phanerochaete chrysosporium*, Water Res. 34 (2000) 4090–4100.

[111] R.S. Bai, T.E. Abraham, Studies on enhancement of Cr (VI) bio sorption by chemically modified biomass of *Rhizopus nigricans*, Water Res. 36 (2002) 1224–1236.

[112] R.K. Dhir, Pulverized Fuel Ash in Cement Replacement Materials in Concrete, Technology and Design, 3, Surrey University Press, London, 1986.

[113] G. Ferraiolo, M. Zilli, A. Converti, Fly ash utilization, J. Chem. Technol. Biotechnol. 47 (1990) 281–305.

[114] M. Grover, M.S. Narayanaswamy, Removal of hexavalent chromium by adsorption on fly ash, J. Inst. Eng., India, Environ. Eng. Div. 1 (1982) 36–39.

[115] W.L. Daniels, B. Stewart, M. Jackson, Utilization of Fly ash to prevent acid mine drainage from coal reuse, in: Proceedings of the 10th International Ash Use Symposium, Vol. 22, American Coal Ash Association and Electric Power Research Institute, Palo Alto, CA, 1993. 1.

[116] K.K. Panday, G. Prasad, V.N. Singh, Mixed adsorbents for Cu(II) removal from aqueous solutions, Environ. Technol. Lett. 7 (1986) 547–554.

[117] A. Kapoor, T. Viraraghavan, Treatment of metal industrial wastewater by fly ash and cement fixation, J. Environ. Eng. 122 (1996) 243–244.

[118] A. Kapoor, T. Viraraghavan, Adsorption of mercury from wastewater by fly ash, Adsorp. Sci. Technol. 9 (1992) 130–147.

[119] T. Viraraghavan, M.M. Dronamraju, Use of fly ash in the removal of copper, nickel and zinc from wastewater, Water Pollut. Res. J. Can. 28 (1993) 369–384.

[120] H.M. Aly, A.A.M. Daifullah, Potential use of bagasse pith for the treatment of wastewater containing metals, Adsorp. Sci. Technol. 16 (1998) 33–38.

[121] A. Papandreou, C.J. Stournaras, D. Panias, Copper and cadmium adsorption on pellets made from fired coal fly ash, J. Hazard. Mater. 148 (2007) 538–547.

[122] C.A. Ríos, C.D. Williams, C.L. Roberts, Removal of heavy metals from acid mine drainage (AMD) using coal fly ash, natural clinker and synthetic zeolites, J. Hazard. Mater. 156 (2008) 23–35.

[123] S. Mohan, R. Gandhimathi, Removal of heavy metal ions from municipal solid waste leachate using coal fly ash as an adsorbent, J. Hazard. Mater. 169 (2009) 351–359.

[124] T. Mishra, S.K. Tiwari, Studies on sorption properties of zeolite derived from Indian fly ash, J. Hazard. Mater. 137 (2006) 299–303.

[125] Z. Elouear, J. Bouzid, N. Boujelben, Removal of nickel and cadmium from aqueous solutions by sewage sludge ash: study in single and binary systems, Environ. Technol. 30 (2009) 561–570.

[126] V.K. Jha, M. Matsuda, M. Miyake, .Sorption properties of the activated carbon-zeolite composite prepared from coal fly ash for Ni(II), Cu(II), Cd(II) and Pb(II), J. Hazard. Mater. 160 (2008) 148–153.

[127] S.K. Srivastava, R. Tyagi, N. Pant, Adsorption of heavy metal ions on carbonaceous material developed from the waste slurry generated in local fertilizer plants, Water Res. 23 (1989) 1161–1169.

[128] S.K. Srivastava, R. Tyagi, N. Pal, D. Mohan, Process development for removal of substituted phenol by carbonaceous adsorbent obtained from fertilizer waste, J. Environ. Eng. 123 (1997) 842–851.

[129] F. Hongxiang, L. Gongxuan, L. Shuben, Adsorption of chromium(VI) ions on to TiO_2 from aqueous solution, Adsorp. Sci. Technol. 16 (1998) 117–120.

[130] M.P. Dahal, G.A. Lawrance, Adsorption of thallium(I), lead(II), copper(II), bismuth(III) and chromium(III) by electrolytic manganese dioxide, Adsorp. Sci. Technol. 13 (1996) 231–240.

[131] J. Liu, S.M. Howard, K.N. Han, Adsorption behavior of cadmium and zinc ions on oxide/water interfaces, Langmuir 9 (1993) 3635–3639.

[132] C. Namasivayam, K. Ranganathan, Effect of organic ligands on the removal of Pb(II), Ni(II) and Cd(II) by 'waste' Fe(III)/Cr(III) hydroxide, Water Res. 32 (1998) 969–971.

[133] C. Namasivayam, S. Senthilkumar, Recycling of industrial wastes: 'Waste' Fe(III)/Cr(III) hydroxide as an adsorbent for the removal of toxic ions and dyes from wastewater, Adsorp. Sci. Technol. 12 (1995) 293–296.

[134] C. Namasivayam, K. Ranganathan, Recycling of 'waste' Fe(III)/Cr (III) hydroxide for the removal of nickel from wastewater: adsorption and equilibrium studies, Waste Manage. 14 (1994) 709–716.

[135] C. Namasivayam, S. Senthilkumar, Recycling of industrial solid waste for the removal of mercury(II) by adsorption process, Chemosphere 34 (1997) 357–375.

[136] S.K. Srivastava, G. Bhattacharjee, R. Tyagi, N. Pant, N. Pal, Studies on the removal of the some toxic metal ions from aqueous solutions and industrial waste. Part (I) (removal of lead and cadmium by hydrous iron and aluminum oxide), Environ. Technol. Lett. 9 (1988) 1173–1185.

[137] R.M. Floroiu, A.P. Davis, A. Torrents, Cadmium adsorption on aluminum oxide in the presence of poly-acrylic acid, Environ. Sci. Technol. 35 (2001) 348–353.

[138] Y.L. Loretta, Properties of red mud tailings produced under varying process conditions, J. Environ. Eng. 124 (1998) 254–264.

[139] E. Lopez, B. Soto, M. Arias, A. Nunez, D. Rubinos, M.T. Barral, Adsorbent properties of red mud and its use for wastewater treatment, Water Res. 32 (1998) 1314–1322.

[140] R. Apak, E. Tutem, M. Hugul, J. Hizal, Heavy metal cation retention by unconventional sorbents (red muds and fly ashes), Water Res. 32 (1998) 430–440.

[141] H.S. Altundoan, S. Altundoan, F. Tümen, M. Bildik, Arsenic removal from aqueous solutions by adsorption on red mud, Waste Manage. 22 (2002) 357–363.

[142] F. Cadena, R. Rizvi, R.W. Peters, Feasibility studies for the removal of heavy metals from solution using tailored Bentonit, in: Hazardous and Industrial Wastes; Proceedings of the Twenty Second Mid Atlantic Industrial Waste Conference, Drexel University, 1990, p. 77.

[143] R.A. Griffin, R.R. Frost, A.K. Au, G.D. Robinson, N.F. Shrimp, Attenuation of pollutants in municipal landfill leachate by clay minerals: heavy metal adsorption, Environ. Geol. Notes 79 (1977) 1–47.

[144] C. Sharma, G.S. Gupta, G. Prasad, D.C. Rupainwar, Use of wollastonite in the removal of Ni (II) from aqueous solutions, Water Air Soil Pollut. 49 (1990) 69–79.

[145] K.K. Panday, G. Prasad, V.N. Singh, Removal of Cr(VI) from aqueous solutions by adsorption on fly ash-wollastonite, J. Chem. Technol. Biotechnol. A 34 (1984) 367–374.

[146] A. Beveridge, W.F. Pickring, The influence of surfactants on the adsorption of heavy metal ions by clays, Water Res. 17 (1983) 215–225.

[147] S.K. Srivastava, R. Tyagi, N. Pant, N. Pal, Studies on the removal of some toxic metal ions. Part II (removal of lead and cadmium by montmorillonite and kaolinite), Environ. Technol. Lett. 10 (1989) 275–282.

[148] K.P. Yadava, B.S. Tyagi, V.N. Singh, Removal of arsenic (III) from aqueous solution by china clay, Environ. Technol. Lett. 9 (1988) 1233–1244.

[149] A.K. Singh, D.P. Singh, V.N. Singh, Removal of Zn(II) from water by adsorption on china clay, Environ. Technol. Lett. 9 (1988) 1153–1162.

[150] S. Hawash, H. El-Abd, M.S. El-Geundi, M.M. Nassar, J.Y. Farah, Useful adsorption equilibriums by means of natural clay, Adsorp. Sci. Technol. 10 (1993) 231–238.

[151] T. Viraraghavan, A. Kapoor, Adsorption of mercury from wastewater by bentonite, Appl. Clay Sci. 9 (1994) 31–49.

[152] D.B. Singh, G.S. Gupta, G. Prasad, D.C. Rupainwaor, Use of hematite for chromium (VI) removal, J. Environ. Sci. Health, Part A, Environ. Sci. Eng. 28 (1993) 1813–1826.

[153] S.A. Khan, R. Rehman, M.A. Khan, Adsorption of chromium(III), chromium(VI) and silver(I) on bentonite, Waste Manage. 15 (1995) 271–282.

[154] A. Mellah, S. Chegrouche, The removal of zinc from aqueous solutions by natural bentonite, Water Res. 31 (1997) 621–629.

[155] Adsorption of cadmium and zinc from aqueous solution on natural and activated bentonite, J. Chem. Technol. Biotechnol. 59 (1994) 289–295.

[156] E.L. Tavani, C. Volzone, Adsorption of chromium(III) from a tanning wastewater on kaolinite, J. Soc. Leather Technol. Chem. 81 (1997) 143–148.

[157] M.J. Angove, B.B. Johnson, J.D. Wells, Adsorption of cadmium(II) on kaolinite, colloids and surfaces, Phys. Chem. Eng. Asp. 126 (1997) 137–147.

[158] D.M. Manohar, K. Anoop Krishnan, T.S. Anirudhan, Removal of mercury (II) from aqueous solutions and chlor-alkali industry wastewater using 2-mercaptobenzimidazole clay, Water Res. 36 (2002) 1609–1619.

[159] Z. Lin, R.W. Puls, Adsorption, desorption and oxidation of arsenic affected by clay minerals and aging process, Environ. Geol. 39 (2000) 753–759.

[160] W.A. Carvalho, C. Vignado, J. Fontana, Ni(II) removal from aqueous effluents by silylated clays, J. Hazard. Mater. 153 (2008) 1240–1247.

[161] Y. Asçi, M. Nurbas, Y.S. Açikel, A comparative study for the sorption of Cd(II) by soils with different clay contents and mineralogy and the recovery of Cd(II) using rhamnolipid biosurfactant, J. Hazard. Mater. 154 (2008) 663–673.

[162] I. Chaari, E. Fakhfakh, S. Chakroun, J. Bouzid, N. Boujelben, M. Feki, F. Rocha, F. Jamoussi, Lead removal from aqueous solutions by a Tunisian smectitic clay, J. Hazard. Mater. 156 (2008) 545–551.

[163] M.Q. Jiang, Q.P. Wang, X.Y. Jin, Z.L. Chen, Removal of Pb(II) from aqueous solution using modified and unmodified kaolinite clay, J. Hazard. Mater. 170 (2009) 332–339.

[164] D.L. Guerra, A.A. Pinto, R.R. Viana, C. Airoldi, Layer silicates modified with 1,4-bis(3-aminopropyl) piperazine for the removal of Th(IV), U(VI) and Eu(III) from aqueous media, J. Hazard. Mater. 171 (2009) 514–523.

[165] M.G. Vieira, A.F. Neto, M.L. Gimenes, M.G. da Silva, Sorption kinetics and equilibrium for the removal of nickel ions from aqueous phase on calcined Bofe bentonite clay, J. Hazard. Mater. 177 (2010) 362–371.

[166] T.S. Anirudhan, C.D. Bringle, S. Rijith, Removal of uranium(VI) from aqueous solutions and nuclear industry effluents using humic acid-immobilized zirconium-pillared clay, J. Environ. Radioact. 101 (2010) 267–276.

[167] E. Nehrenheim, J.P. Gustafsson, Kinetic sorption modelling of Cu(II), Ni(II), Zn(II), Pb(II) and Cr(III) ions to pine bark and blast furnace slag by using batch experiments, Bioresour. Technol. 99 (2008) 1571–1577.

[168] D.W. Ming, J.B. Dixon, Quantitative determination of clinoptilolite in soils by a cation-exchange capacity method, Clays Clay Miner. 35 (1987) 463–468.

[169] D. Leppert, Heavy metal sorption with clinoptilolite zeolite: alternatives for treating contaminated soil and water, Min. Eng. 42 (1990) 604–608.

[170] I. Santiago, V.P. Worland, E. Cazares-Rivera, F. Cadena, Adsorption of Hexavalent chromium onto tailored zeolites, in: 47th Purdue Industrial Waste Conference Proceedings, Lewis Publishers, Inc., Chelsea, MI, 1992, p. 669.

[171] A.A.H. Ali, R. El-Bishtawi, Removal of lead and nickel ions using zeolite tuff, J. Chem. Technol. Biotechnol. 69 (1997) 27–34.

[172] M.I. Panayotoya, Kinetics and thermodynamics of copper ions removal from wastewater by use of zeolite, Waste Manage. 21 (2001) 671–676.

[173] J.S. Kim, L. Zhag, M.A. Keane, Iron removal from aqueous solutions by batch ion exchange with a solid Na – Y zeolite, Sep. Sci. Technol. 36 (2001) 1509–1525.

[174] D.T. Bostick, S.M. DePaoli, B. Guo, The removal of 90Sr and 137Cs by natural chabazite zeolite, Sep. Sci. Technol. 36 (2001) 975–998.

[175] H.M. Baker, A.M. Massadeh, H.A. Younes, Natural Jordanian zeolite: removal of heavy metal ions from water samples using column and batch methods, Environ. Monit. Assess. 157 (2009) 319–330.

[176] S. Khaodhiar, M.F. Azizian, K. Osathaphan, P.O. Nelson, Copper, chromium, and arsenic adsorption and equilibrium modeling in an iron oxide coated sand, background electrolyte system, Water Air Soil Pollut. 119 (2000) 105–120.

[177] S.L. Huang, Cadmium adsorption by sediment in a turbulence tank, Water Res. 35 (2001) 2635–2644.

[178] N. Dali-Youcef, B. Ouddane, Z. Derriche, Adsorption of zinc on natural sediment of Tafna River (Algeria), J. Hazard. Mater. 137 (2006) 1263–1270.

[179] M. Murakami, F. Nakajima, H. Furumai, The sorption of heavy metal species by sediments in soak ways receiving urban road runoff, Chemosphere 70 (2008) 2099–2109.

[180] K. Flogeac, E. Guillon, M. Aplincourt, Adsorption of several metal ions onto a model soil sample: equilibrium and EPR studies, J. Colloid Interface Sci. 286 (2005) 596–601.

[181] Y.J. Wang, D.M. Zhou, R.J. Sun, L. Cang, X.Z. Hao, Cosorption of zinc and glyphosate on two soils with different characteristics, J. Hazard. Mater. 137 (2006) 76–82.

[182] S. Chakravarty, V. Dureja, G. Bhattacharyya, S. Maity, S. Bhattacharjee, Removal of arsenic from groundwater using low cost ferruginous manganese ore, Water Res. 36 (2002) 625–632.

[183] J.D. Prikryl, A. Jain, D.R. Turner, R.T. Pabalan, Uranium(VI) sorption behavior on silicate mineral mixtures, J. Contam. Hydrol. 47 (2001) 241–253.

[184] A.G. Leyva, J. Marrero, P. Smichowski, D. Cicerone, Sorption of antimony onto hydroxyapatite, Environ. Sci. Technol. 35 (2001) 3669–3675.

[185] C. Papelis, C. Papelis, Cation and anion sorption on granite from the Project Shoal Test Area, near Fallon, Nevada, USA, Adv. Environ. Res. 5 (2001) 151–166.

[186] H. Ghassabzadeh, A. Mohadespour, M. Torab-Mostaedi, P. Zaheri, M.G. Maragheh, H. Taheri, Adsorption of Ag, Cu and Hg from aqueous solutions using expanded perlite, J. Hazard. Mater. 177 (2010) 950–955.

[187] M.E. Argun, S. Dursun, A new approach to modification of natural adsorbent for heavy metal adsorption, Bioresour. Technol. 99 (2008) 2516–2527.

[188] P. Ahuja, R. Gupta, R.K. Saxena, Sorption and desorption of cobalt by Oscillatoria angustissima, Curr. Microbiol. 39 (1999) 49–52.

[189] H.J. Zhu, Y.F. Jia, X. Wu, H. Wang, Removal of arsenite from drinking water by activated carbon supported nano zero-valent iron, Huan. Jing. Ke. Xue. 30 (2009) 1644–1648.

[190] U. Garg, M.P. Kaur, G.K. Jawa, D. Sud, V.K. Garg, Removal of cadmium (II) from aqueous solutions by adsorption on agricultural waste biomass, J. Hazard. Mater. 154 (2008) 1149–1157.

[191] M. Tuzen, A. Sari, D. Mendil, O.D. Uluozlu, M. Soylak, M. Dogan, Characterization of biosorption process of As(III) on green algae Ulothrix cylindricum, J. Hazard. Mater. 165 (2009) 566–572.

[192] A. Corami, S. Mignardi, V. Ferrini, Cadmium removal from single- and multi-metal (Cd^+ Pb^+ Zn^+ Cu^+) solutions by sorption on hydroxyapatite, J. Colloid Interface Sci. 317 (2008) 402–408.

[193] G.S. Murugesan, M. Sathishkumar, K. Swaminathan, Arsenic removal from groundwater by pretreated waste tea fungal biomass, Bioresour. Technol. 3 (2006) 483–487.

[194] A. Ramesh, H. Hasegawa, W. Sugimoto, T. Maki, K. Ueda, Adsorption of gold(III), platinum(IV) and palladium(II) onto glycine modified crosslinked chitosan resin, Bioresour. Technol. 99 (2008) 3801–3809.

[195] C.R. Adhikari, D. Parajuli, H. Kawakita, K. Inoue, K. Ohto, H. Harada, Dimethylamine-modified waste paper for the recovery of precious metals, Environ. Sci. Technol. 42 (2008) 5486–5491.

[196] L. Chai, Y. Chen, Z. Yang, Kinetics and thermodynamics of arsenate and arsenite biosorption by pretreated spent grains, Water Environ. Res. 81 (2009) 843–848.

[197] Y. Qing, Y. Hang, R. Wanjaul, Z. Jiang, B. Hu, Adsorption behavior of noble metal ions (Au, Ag, Pd) on nanometer-size titanium dioxide with ICP-AES, Anal. Sci. 19 (2003) 1417–1420.

[198] E. Pehlivan, B.H. Yanik, G. Ahmetli, M. Pehlivan, Equilibrium isotherm studies for the uptake of cadmium and lead ions onto sugar beet pulp, Bioresour. Technol. 99 (2008) 3520–3527.

[199] A. Sari, M. Tuzen, Biosorption of As(III) and As(V) from aqueous solution by macrofungus (Inonotus hispidus) biomass: equilibrium and kinetic studies, J. Hazard. Mater. 164 (2009) 1372–1378.

[200] A.B. Farag, M.H. Soliman, O.S. Abdel-Rasoul, M.S. el-Shahawi, Sorption characteristics and chromatographic separation of gold (I and III) from silver and base metal ions using polyurethane foams, Anal. Chim. Acta. 601 (2007) 218–229.

[201] G. Chen, G. Zeng, L. Tang, C. Du, X. Jiang, G. Huang, H. Liu, G. Shen, Cadmium removal from simulated wastewater to biomass byproduct of Lentinus edodes, Bioresour. Technol. 99 (2008) 7034–7040.

[202] P.K. Pandey, Y. Verma, S. Choubey, M. Pandey, K. Chandrasekhar, Biosorptive removal of cadmium from contaminated groundwater and industrial effluents, Bioresour. Technol. 99 (2008) 4420–4427.

[203] M. Mohapatra, S.K. Sahoo, S. Anand, R.P. Das, Removal of As(V) by Cu(II), Ni(II), or Co(II)-doped goethite samples, J. Colloid Interface Sci. 298 (2006) 6–12.

[204] S.B. Choi, Y.S. Yun, Biosorption of cadmium by various types of dried sludge: an equilibrium study and investigation of mechanisms, J. Hazard. Mater. 138 (2006) 378–383.

[205] Y. Liu, Q. Cao, F. Luo, J. Chen, Biosorption of Cd(II), Cu(II), Ni(II) and Zn(II) ions from aqueous solutions by pretreated biomass of brown algae, J. Hazard. Mater. 163 (2009) 931–938.

[206] P. Lodeiro, C. Rey-Castro, J.L. Barriada, M.E. Sastre de Vicente, R. Herrero, Biosorption of cadmium by the protonated macroalga *Sargassum muticum*: binding analysis with a nonideal, competitive, and thermodynamically consistent adsorption (NICCA) model, J. Colloid Interface Sci. 289 (2005) 352–358.

[207] Y. Göksungur, S. Uren, U. Güvenç, Biosorption of cadmium and lead ions by ethanol treated waste baker's yeast biomass, Bioresour. Technol. 96 (2005) 103–109.

[208] R.J. Martins, R. Pardo, R.A. Boaventura, Cadmium(II) and zinc(II) adsorption by the aquatic moss Fontinalis antipyretica: effect of temperature, pH and water hardness, Water Res. 38 (2004) 693–699.

[209] V.C. Taty-Costodes, H. Fauduet, C. Porte, A. Delacroix, Removal of Cd(II) and Pb(II) ions, from aqueous solutions, by adsorption onto sawdust of *Pinus sylvestris*, J. Hazard. Mater. 105 (2003) 121–142.

[210] M. Luisa Cervera, M. Carmen Arnal, M. de la Guardia, Removal of heavy metals by using adsorption on alumina or chitosan, Anal. Bioanal. Chem. 375 (2003) 820–825.

[211] A.A. Abia, M. Horsfall Jr., O. Didi, The use of chemically modified and unmodified cassava waste for the removal of Cd, Cu and Zn ions from aqueous solution, Bioresour. Technol. 90 (2003) 345–348.

[212] M. Iqbal, A. Saeed, Removal of heavy metals from contaminated water by petiolar felt-sheath of palm, Environ. Technol. 23 (2002) 1091–1098.

[213] J. Marin, J. Ayele, Removal of some heavy metal cations from aqueous solutions by spruce sawdust. I. Study of the binding mechanism through batch experiments, Environ. Technol. 23 (2002) 1157–1171.

[214] A. Benhammou, A. Yaacoubi, L. Nibou, B. Tanouti, Adsorption of metal ions onto Moroccan stevensite: kinetic and isotherm studies, J. Colloid Interface Sci. 282 (2005) 320–326.

[215] S. Saygideger, O. Gulnaz, E.S. Istifli, N. Yucel, Adsorption of Cd(II), Cu(II) and Ni(II) ions by Lemna minor L.: effect of physicochemical environment, J. Hazard. Mater. 126 (2005) 96–104.

[216] Z. Li, H. Yuan, Characterization of cadmium removal by Rhodotorula sp. Y11, Appl. Microbiol. Biotechnol. 73 (2006) 458–463.

[217] Y. Prasanna Kumar, P. King, V.S. Prasad, Equilibrium and kinetic studies for the biosorption system of copper(II) ion from aqueous solution using Tectona grandis L.f. leaves powder, J. Hazard. Mater. 137 (2006) 1211–1217.

[218] R. Vimala, N. Das, Biosorption of cadmium(II) and lead(II) from aqueous solutions using mushrooms: a comparative study, J. Hazard. Mater. 168 (2009) 376–382.

[219] V.K. Verma, S. Tewari, J.P. Rai, Ion exchange during heavy metal bio-sorption from aqueous solution by dried biomass of macrophytes, Bioresour. Technol. 99 (2008) 1932–1938.

[220] A. Agrawal, K.K. Sahu, Kinetic and isotherm studies of cadmium adsorption on manganese nodule residue, J. Hazard. Mater. 137 (2006) 915–924.

[221] M.S. Al-Masri, Y. Amin, B. Al-Akel, T. Al-Naama, Biosorption of cadmium, lead, and uranium by powder of poplar leaves and branches, Appl. Biochem. Biotechnol. 160 (2010) 976–987.

[222] T.K. Naiya, A.K. Bhattacharya, S.K. Das, Adsorption of Cd(II) and Pb(II) from aqueous solutions on activated alumina, J. Colloid. Interface Sci. 333 (2009) 14–26.

[223] A. Sari, M. Tuzen, Biosorption of cadmium(II) from aqueous solution by red algae (Ceramium virgatum): equilibrium, kinetic and thermodynamic studies, J. Hazard. Mater. 157 (2008) 448–454.

[224] R. Salim, M. Al-Subu, E. Dawod, Efficiency of removal of cadmium from aqueous solutions by plant leaves and the effects of interaction of combinations of leaves on their removal efficiency, J. Environ. Manage. 87 (2008) 521–532.

[225] C.W. Cheung, J.F. Porter, G. Mckay, Sorption kinetic analysis for the removal of cadmium ions from effluents using bone char, Water Res. 35 (2001) 605–612.

[226] W. Zheng, X.M. Li, Q. Yang, G.M. Zeng, X.X. Shen, Y. Zhang, J.J. Liu, Adsorption of Cd(II) and Cu(II) from aqueous solution by carbonate hydroxylapatite derived from eggshell waste, J. Hazard. Mater. 147 (2007) 534–539.

[227] N.V. Farinella, G.D. Matos, E.L. Lehmann, M.A. Arruda, Grape bagasse as an alternative natural adsorbent of cadmium and lead for effluent treatment, J. Hazard. Mater. 154 (2008) 1007–1012.

[228] M. Terashima, N. Oka, T. Sei, H. Yoshida, Adsorption of cadmium ion and gallium ion to immobilized metallothionein fusion protein, Biotechnol. Prog. 8 (2002) 1318–1323.

[229] D.K. Singh, S. Mishra, Synthesis, characterization and removal of Cd(II) using Cd(II) ion imprinted polymer, J. Hazard. Mater. 164 (2009) 1547–1551.

[230] G. Carrillo-Morales, M.M. Dávila-Jiménez, M.P. Elizalde-González, A.A. Peláez-Cid, Removal of metal ions from aqueous solution by adsorption on the natural adsorbent CACMM2, J. Chromatogr. A 938 (2001) 237–242.

[231] G.T. Tsao, Y. Zheng, J. Lu, C.S. Gong, Adsorption of heavy metal ions by immobilized phytic acid, Appl. Biochem. Biotechnol. 63–65 (1997) 731–741.

[232] B. Salih, A. Denizli, C. Kavaklı, R. Say, E. Piskin, Adsorption of heavy metal ions onto dithizone-anchored poly (EGDMA-HEMA) microbeads, Talanta 46 (1998) 1205–1213.

[233] X. Huang, X. Liao, B. Shi, Tannin-immobilized mesoporous silica bead (BT-SiO2) as an effective adsorbent of Cr(III) in aqueous solutions, J. Hazard. Mater. 173 (2010) 33–39.

[234] B. Jha, S. Basha, S. Jaiswar, B. Mishra, M.C. Thakur, Biosorption of Cd(II) and Pb(II) onto brown seaweed, Lobophora variegata (Lamouroux): kinetic and equilibrium studies, Biodegradation 20 (2009) 1–13.

[235] O.M. Freitas, R.J. Martins, C.M. Delerue-Matos, R.A. Boaventura, Removal of Cd(II), Zn(II) and Pb(II) from aqueous solutions by brown marine macro algae: kinetic modeling, J. Hazard. Mater. 153 (2008) 493–501.

[236] V.T. Vinod, R.B. Sashidhar, A.A. Sukumar, Competitive adsorption of toxic heavy metal contaminants by gum kondagogu (*Cochlospermum gossypium*): a natural hydrocolloid, Colloids Surf. B Biointerfaces 75 (2010) 490–495.

[237] M.E. Argun, S. Dursun, M. Karatas, M. Gürü, Activation of pine cone using Fenton oxidation for Cd(II) and Pb(II) removal, Bioresour. Technol. 99 (2008) 8691–8698.

[238] R. Chand, K. Narimura, H. Kawakita, K. Ohto, T. Watari, K. Inoue, Grape waste as a biosorbent for removing Cr(VI) from aqueous solution, J. Hazard. Mater. 163 (2009) 245–250.

[239] J.R. Memon, S.Q. Memon, M.I. Bhanger, A. El-Turki, K.R. Hallam, G.C. Allen, Banana peel: a green and economical sorbent for the selective removal of Cr(VI) from industrial wastewater, Colloids. Surf. B Biointerfaces 70 (2009) 232–237.

[240] G.R. Bernardo, R.M. Rene, A.D. Ma Catalina, Chromium(III) uptake by agro-waste biosorbents: chemical characterization, sorption-desorption studies, and mechanism, J. Hazard. Mater. 170 (2009) 845–854.

[241] N.R. Bishnoi, R. Kumar, K. Bishnoi, Biosorption of Cr(VI) with Trichoderma viride immobilized fungal biomass and cell free Ca-alginate beads, Indian J. Exp. Biol. 45 (2007) 657–664.

[242] N. Ertugay, Y.K. Bayhan, Biosorption of Cr(VI) from aqueous solutions by biomass of Agaricus bisporus, J. Hazard. Mater. 154 (2008) 432–439.

[243] B.S. Inbaraj, J.S. Wang, J.F. Lu, F.Y. Siao, B.H. Chen, Adsorption of toxic mercury(II) by an extracellular biopolymer poly(gamma-glutamic acid), Bioresour. Technol. 100 (2009) 200–207.

[244] M. Jain, V.K. Garg, K. Kadirvelu, Adsorption of hexavalent chromium from aqueous medium onto carbonaceous adsorbents prepared from waste biomass, J. Environ. Manage. 91 (2010) 949–957.

[245] Z. Aksu, E. Balibek, Chromium(VI) biosorption by dried Rhizopus arrhizus: effect of salt (NaCl) concentration on equilibrium and kinetic parameters, J. Hazard. Mater. 145 (2007) 210–220.

[246] J. Wang, D. Zhang, T.R. Lawson, R.A. Bartsch, Sorption of heavy metal ions by silica gel-immobilized, proton-ionizable calix[4]arenes, Talanta 78 (2009) 477–483.

[247] F. Gode, E.D. Atalay, E. Pehlivan, Removal of Cr(VI) from aqueous solutions using modified red pine sawdust, J. Hazard. Mater. 152 (2008) 1201–1207.

[248] J. Hu, G. Chen, I.M. Lo, Removal and recovery of Cr(VI) from wastewater by maghemite nanoparticles, Water Res. 39 (2005) 4528–4536.

[249] S.H. Baysal, S. Onal, G. Ozdemir, Biosorption of chromium, cadmium, and cobalt from aqueous solution by immobilized living cells of Chryseomonas luteola TEM 05, Prep. Biochem. Biotechnol. 39 (2009) 419–428.

[250] N.R. Bishnoi, R. Kumar, S. Kumar, S. Rani, Biosorption of Cr(III) from aqueous solution using algal biomass spirogyra spp, J. Hazard. Mater. 145 (2007) 142–227.

[251] N.R. Bishnoi, M. Bajaj, N. Sharma, Adsorption of Cr(VI) from aqueous and electroplating wastewater, Environ. Technol. 25 (2004) 899–905.

[252] K. Chojnacka, Biosorption of Cr(III) ions by eggshells, J. Hazard. Mater. 121 (2005) 167–173.

[253] A.V. Bankar, A.R. Kumar, S.S. Zinjarde, Removal of chromium (VI) ions from aqueous solution by adsorption onto two marine isolates of Yarrowia lipolytica, J. Hazard. Mater. 170 (2009) 487–494.

[254] J.S. Melo, S.F. D'Souza, Removal of chromium by mucilaginous seeds of Ocimum basilicum, Bioresour. Technol. 92 (2004) 151–155.

[255] P.L. Tang, C.K. Lee, K.S. Low, Z. Zainal, Sorption of Cr(VI) and Cu(II) in aqueous solution by ethylenediamine modified rice hull, Environ. Technol. 24 (2003) 1243–1251.

[256] M. Khamis, F. Jumean, N. Abdo, Speciation and removal of chromium from aqueous solution by white, yellow and red UAE sand, J. Hazard. Mater. 169 (2009) 948–952.

[257] R.S. Bai, T.E. Abraham, Biosorption of Cr(VI) from aqueous solution by *Rhizopus nigricans*, Bioresour. Technol. 79 (2001) 73–81.

[258] M.E. Mahmoud, O.F. Hafez, M.M. Osman, A.A. Yakout, A. Alrefaay, Hybrid inorganic/organic alumina adsorbents-functionalized-purpurogallin for removal and preconcentration of Cr(III), Fe(III), Cu(II), Cd(II) and Pb(II) from underground water, J. Hazard. Mater. 176 (2010) 906–912.

[259] J. Yao, H. Xu, J. Wang, M. Jiang, P. Ouyang, Removal of Cr(III), Ni(II) and Cu(II) by poly(gamma-glutamic acid) from Bacillus subtilis NX-2, J. Biomater. Sci. Polym. Ed. 18 (2007) 193–204.

[260] R.M. Pérez Silva, A. Abalos Rodríguez, J.M. Gómez Montes De Oca, D. Cantero Moreno, Biosorption of chromium, copper, manganese and zinc by *Pseudomonas aeruginosa* AT18 isolated from a site contaminated with petroleum, Bioresour. Technol. 100 (2009) 1533–1538.

[261] P. Yin, Q. Xu, R. Qu, G. Zhao, Y. Sun, Adsorption of transition metal ions from aqueous solutions onto a novel silica gel matrix inorganic-organic composite material, J. Hazard. Mater. 173 (2010) 710–716.

[262] M. Yiğitoğlu, M. Arslan, Selective removal of Cr(VI) ions from aqueous solutions including Cr(VI), Cu(II) and Cd(II) ions by 4-vinyl pyridine/2-hydroxyethylmethacrylate monomer mixture grafted poly(ethylene terephthalate) fiber, J. Hazard. Mater. 166 (2009) 435–444.

[263] I.G. Shibi, T.S. Anirudhan, Adsorption of Co(II) by a carboxylate-functionalized polyacrylamide grafted lignocellulosics, Chemosphere 58 (2005) 1117–1126.

[264] A.K. Meena, K. Kadirvelu, G.K. Mishra, C. Rajagopal, P.N. Nagar, Adsorptive removal of heavy metals from aqueous solution by treated sawdust (*Acacia arabica*), J. Hazard. Mater. 150 (2008) 604–611.

[265] Y.H. Wang, S.H. Lin, R.S. Juang, Removal of heavy metal ions from aqueous solutions using various low-cost adsorbents, J. Hazard. Mater. 102 (2003) 291–302.

[266] R. Vannela, S.K. Verma, Co(II), Cu(II), and Zn(II) accumulation by cyanobacterium *Spirulina platensis*, Biotechnol. Prog. 22 (2006) 1282–1293.

[267] R. Jalali-Rad, H. Ghafourian, Y. Asef, S.T. Dalir, M.H. Sahafipour, B.M. Gharanjik, Biosorption of cesium by native and chemically modified biomass of marine algae: introduce the new biosorbents for biotechnology applications, J. Hazard. Mater. 116 (2004) 125–134.

[268] C.H. Hsieh, S.L. Lo, W.H. Kuan, C.L. Chen, Adsorption of copper ions onto microwave stabilized heavy metal sludge, J. Hazard. Mater. 136 (2006) 338–344.

[269] M.L. Schlegel, L. Charlet, A. Manceau, Sorption of metal ions on clay minerals, J. Colloid. Interface Sci. 220 (1999) 392–405.

[270] Z. Yu, R. Wu, M. Wu, L. Zhao, R. Li, H. Zou, Preparation of polyamine-functionalized copper specific adsorbents for selective adsorption of copper, Colloids Surf. B Biointerfaces. 78 (2010) 222–228.

[271] V.J. Vilar, C.M. Botelho, J.M. Loureiro, R.A. Boaventura, Biosorption of copper by marine algae Gelidium and algal composite material in a packed bed column, Bioresour. Technol. 99 (2008) 5830–5838.

[272] D. Karamanis, P.A. Assimakopoulos, Efficiency of aluminum-pillared montmorillonite on the removal of cesium and copper from aqueous solutions, Water Res. 41 (2007) 1897–1906.

[273] H.D. Ozsoy, H. Kumbur, B. Saha, J.H. van Leeuwen, Use of Rhizopus oligosporus produced from food processing wastewater as a biosorbent for Cu(II) ions removal from the aqueous solutions, Bioresour. Technol. 99 (2008) 4943–4948.

[274] B. Kannamba, K.L. Reddy, B.V. Appa Rao, Removal of Cu(II) from aqueous solutions using chemically modified chitosan, J. Hazard. Mater. 175 (2010) 939–948.

[275] J. Bouzid, Z. Elouear, M. Ksibi, M. Feki, A. Montiel, A study on removal characteristics of copper from aqueous solution by sewage sludge and pomace ashes, J. Hazard. Mater. 152 (2008) 838–845.

[276] C.Y. Yin, M.K. Aroua, W.M. Daud, Modification of granular activated carbon using low molecular weight polymer for enhanced removal of Cu(II) from aqueous solution, Water Sci. Technol. 56 (2007) 95–101.

[277] H. Yazici, M. Kiliç, M. Solak, Biosorption of copper(II) by Marrubium globosum subsp. globosum leaves powder: effect of chemical pretreatment, J. Hazard. Mater. 151 (2008) 669–675.

[278] B. Zhu, T. Fan, D. Zhang, Adsorption of copper ions from aqueous solution by citric acid modified soybean straw, J. Hazard. Mater. 153 (2008) 300–308.

[279] A. Celekli, M. Yavuzatmaca, H. Bozkurt, An eco-friendly process: predictive modelling of copper adsorption from aqueous solution on Spirulina platensis, J. Hazard. Mater. 173 (2010) 123–129.

[280] S.K. Papageorgiou, F.K. Katsaros, E.P. Kouvelos, J.W. Nolan, H. Le Deit, N.K. Kanellopoulos, Heavy metal sorption by calcium alginate beads from *Laminaria digitata*, J. Hazard. Mater. 137 (2006) 1765–1772.

[281] K. Tsekova, M. Ianis, V. Dencheva, S. Ganeva, Biosorption of binary mixtures of copper and cobalt by Penicillium brevicompactum, Z. Naturforsch C. 62 (2007) 261–264.

[282] L. Deng, X. Zhu, X. Wang, Y. Su, H. Su, Biosorption of copper(II) from aqueous solutions by green alga Cladophora fascicularis, Biodegradation 18 (2007) 393–402.

[283] S.S. Mark, T.C. Crusberg, C.M. Dacunha, A.A. Di Iorio, A heavy metal biotrap for wastewater remediation using poly-gamma-glutamic acid, Biotechnol. Prog. 22 (2006) 523–531.

[284] R. Han, Z. Lu, W. Zou, W. Daotong, J. Shi, Y. Jiujun, Removal of copper(II) and lead(II) from aqueous solution by manganese oxide coated sand II. Equilibrium study and competitive adsorption, J. Hazard. Mater. 137 (2006) 480–488.

[285] J.K. Yang, M.R. Yu, S.M. Lee, Removal of toxic heavy metals by iron-coated starfish, Water Sci. Technol. 56 (2007) 51–57.

[286] H.D. Ozsoy, H. Kumbur, Adsorption of Cu(II) ions on cotton ball, J. Hazard. Mater. 136 (2006) 911–916.

[287] R. Han, J. Zhang, W. Zou, H. Xiao, J. Shi, H. Liu, Biosorption of copper(II) and lead(II) from aqueous solution by chaff in a fixed-bed column, J. Hazard. Mater. 133 (2006) 262–268.

[288] F. Ekmekyapar, A. Aslan, Y.K. Bayhan, A. Cakici, Biosorption of copper(II) by nonliving lichen biomass of Cladonia rangiformis hoffm, J. Hazard. Mater. 137 (2006) 293–298.

[289] P. Kaewsarn, Biosorption of copper(II) from aqueous solutions by pre-treated biomass of marine algae Padina sp, Chemosphere 47 (10) (2002) 1081–1085.

[290] A. Ozcan, A.S. Ozcan, S. Tunali, T. Akar, I. Kiran, Determination of the equilibrium, kinetic and thermodynamic parameters of adsorption of copper(II) ions onto seeds of Capsicum annuum, J. Hazard. Mater. 124 (2005) 200–208.

[291] M.M. Rao, A. Ramesh, G.P. Rao, K. Seshaiah, Removal of copper and cadmium from the aqueous solutions by activated carbon derived from Ceiba pentandra hulls, J. Hazard. Mater. 129 (2006) 123–129.

[292] A. Hassen, F. Jamoussi, N. Saidi, Z. Mabrouki, E. Fakhfakh, Microbial and copper adsorption by smectitic clay – an experimental study, Environ, Technol. 24 (2003) 1117–1127.

[293] S.S. Banerjee, D.H. Chen, Fast removal of copper ions by gum arabic modified magnetic nano-adsorbent, J. Hazard. Mater. 147 (2007) 792–799.

[294] C.H. Lai, C.Y. Chen, Removal of metal ions and humic acid from water by iron-coated filter media, Chemosphere 44 (2001) 1177–1184.

[295] O. Gulnaz, S. Saygideger, E. Kusvuran, Study of Cu(II) biosorption by dried activated sludge: effect of physico-chemical environment and kinetics study, J. Hazard. Mater. 120 (2005) 193–200.

[296] A. Jang, Y. Seo, P.L. Bishop, The removal of heavy metals in urban runoff by sorption on mulch, Environ. Pollut. 133 (2005) 117–127.

[297] D. Bozić, V. Stanković, M. Gorgievski, G. Bogdanović, R. Kovacević, Adsorption of heavy metal ions by sawdust of deciduous trees, J. Hazard. Mater. 171 (2009) 684–692.

[298] S.G. Zhou, L.X. Zhou, F.X. Chen, Characterization and heavy metal adsorption properties of schwertmannite synthesized by bacterial oxidation of ferrous sulfate solutions, Guang. Pu. Xue. Yu. Guang. Pu. Fen. Xi. 27 (2007) 367–370.

[299] K.K. Wong, C.K. Lee, K.S. Low, M.J. Haron, Removal of Cu(II) and Pb(II) by tartaric acid modified rice husk from aqueous solutions, Chemosphere 50 (2003) 23–28.

[300] A. Hammaini, F. González, A. Ballester, M.L. Blázquez, J.A. Muñoz, Biosorption of heavy metals by activated sludge and their desorption characteristics, J. Environ. Manage. 84 (2007) 419–426.

[301] H. Djati Utomo, K.A. Hunter, Adsorption of divalent copper, zinc, cadmium and lead ions from aqueous solution by waste tea and coffee adsorbents, Environ. Technol. 27 (2006) 25–32.

[302] K.C. Bhainsa, S.F. D'Souza, Removal of copper ions by the filamentous fungus, Rhizopus oryzae from aqueous solution, Bioresour. Technol. 99 (2008) 3829–3835.

[303] A. Ozverdi, M. Erdem, Cu(II), Cd(II) and Pb(II) adsorption from aqueous solutions by pyrite and synthetic iron sulphide, J. Hazard. Mater. 137 (2006) 626–632.

[304] Y. Li, Q. Yue, B. Gao, Adsorption kinetics and desorption of Cu(II) and Zn(II) from aqueous solution onto humic acid, J. Hazard. Mater. 178 (2010) 455–461.

[305] C. Niu, W. Wu, Z. Wang, S. Li, J. Wang, Adsorption of heavy metal ions from aqueous solution by crosslinked carboxymethyl konjac glucomannan, J. Hazard. Mater. 141 (2007) 209–214.

[306] R. Han, H. Li, Y. Li, J. Zhang, H. Xiao, J. Shi, Biosorption of copper and lead ions by waste beer yeast, J. Hazard. Mater. 137 (2006) 1569–1576.

[307] A. Sari, M. Tuzen, D. Citak, M. Soylak, Adsorption characteristics of Cu(II) and Pb(II) onto expanded perlite from aqueous solution, J. Hazard. Mater. 148 (2007) 387–394.

[308] X. Chen, W.H. Hou, Q.H. Wang, Removal of metal ions Cu(II), Cd(II) and Pb(II) from solutions by sorption on slag, Huan. Jing. Ke. Xue. 30 (2009) 2940–2945.

[309] R. Gao, Z. Hu, X. Chang, Q. He, L. Zhang, Z. Tu, J. Shi, Chemically modified activated carbon with 1-acylthiosemicarbazide for selective solid-phase extraction and preconcentration of trace Cu(II), Hg(II) and Pb(II) from water samples, J. Hazard. Mater. 172 (2009) 324–329.

[310] M.E. Argun, S. Dursun, C. Ozdemir, M. Karatas, Heavy metal adsorption by modified oak sawdust: thermodynamics and kinetics, J. Hazard. Mater. 141 (2007) 77–85.

[311] D. Xinjiao, Biosorption of Cu^{2+} from aqueous solutions by pretreated Cladosporium sp. J. Environ. Biol. 27 (2006) 639–643.

[312] S. Ahmady-Asbchin, Y. Andres, C. Gerente, P. Le Cloirec, Natural seaweed waste as sorbent for heavy metal removal from solution, Environ. Technol. 30 (2009) 755–762.

[313] B. Nedumaran, M. Velan, Removal of copper(II) ions from aqueous solutions by Azolla rongpong: batch and continuous study, J. Environ. Sci. Eng. 50 (2008) 23–28.

[314] K. Jiang, T.H. Sun, L.N. Sun, H.B. Li, Adsorption characteristics of copper, lead, zinc and cadmium ions by tourmaline, J. Environ. Sci. (China) 18 (2006) 1221–1225.

[315] H.H. Omar, Biosorption of copper, nickel and manganese using non-living biomass of marine alga, *Ulva lactuca*, Pak. J. Biol. Sci. 11 (2008) 964–973.

[316] M. Rafatullah, O. Sulaiman, R. Hashimm, A. Ahmad, Adsorption of copper(II), chromium (III), nickel(II) and lead(II) ions from aqueous solutions by meranti sawdust, J. Hazard. Mater. 170 (2009) 969–977.

[317] M.V. Dinu, E.S. Dragan, Evaluation of Cu(II), Co(II) and Ni(II) ions removal from aqueous solution using a novel chitosan/clinoptilolite composite: kinetics and isotherms, Chem. Eng. J. 160 (2010) 157–163.

[318] E. Pehlivan, A.M. Ozkan, S. Dinç, S. Parlayici, Adsorption of Cu(II) and Pb(II) ion on dolomite powder, J. Hazard. Mater. 167 (2009) 1044–1049.

[319] N. Georgieva, R. Bryaskova, A. Debuigne, C. Detrembleur, Cadmium and copper absorption mediated by a poly(vinyl alcohol)-β-polyacrylonitrile based micelle/*Trichosporon cutaneum* cell system, J. Appl. Poly. Sci. 116 (2010) 2970–2975.

[320] K. Vijayaraghavan, T.T. Teo, R. Balasubramanian, U.M. Joshi, Application of Sargassum biomass to remove heavy metal ions from synthetic multi-metal solutions and urban storm water runoff, J. Hazard. Mater. 164 (2009) 1019–1023.

[321] S.H. Huang, D.H. Chen, Rapid removal of heavy metal cations and anions from aqueous solutions by an amino-functionalized magnetic nano-adsorbent, J. Hazard. Mater. 163 (2009) 174–179.

[322] B. Saha, S. Chakraborty, G. Das, A comparative metal ion adsorption study by trimesic acid coated alumina: a potent adsorbent, J. Colloid Interface Sci. 323 (2008) 26–32.

[323] D. Mohan, S. Chander, Single, binary, and multicomponent sorption of iron and manganese on lignite, J. Colloid Interface Sci. 299 (2006) 76–87.

[324] E. Chockalingam, S. Subramanian, Studies on removal of metal ions and sulphate reduction using rice husk and Desulfotomaculum nigrificans with reference to remediation of acid mine drainage, Chemosphere 62 (2006) 699–708.

[325] C. Green-Ruiz, Mercury(II) removal from aqueous solutions by nonviable Bacillus sp. from a tropical estuary, Bioresour. Technol. 97 (2006) 1907–1911.

[326] S.K. Das, A.R. Das, A.K. Guha, A study on the adsorption mechanism of mercury on Aspergillus versicolor biomass, Environ. Sci. Technol. 41 (2007) 8281–8287.

[327] I. Tüzün, G. Bayramoğlu, E. Yalçin, G. Basaran, G. Celik, M.Y. Arica, Equilibrium and kinetic studies on biosorption of Hg(II), Cd(II) and Pb(II) ions onto microalgae Chlamydomonas reinhardtii, J. Environ. Manage. 77 (2005) 85–92.

[328] Noghabi, K.A., Zahiri, H.S., Lotfi, A.S., Raheb, J., Nasri, S. and Yoon, S.C., Mercury absorption by Pseudomonas fluorescens BM07 grown at two different temperatures, Pol. J. Microbiol. 56 (2007) 111–117.

[329] C. Jeon, K.H. Park, Adsorption and desorption characteristics of mercury(II) ions using aminated chitosan bead, Water Res. 39 (2005) 3938–3944.

[330] A. Oztürk, Removal of nickel from aqueous solution by the bacterium Bacillus thuringiensis, J. Hazard. Mater. 147 (2007) 518–523.

[331] K.G. Bhattacharyya, J. Sarma, A. Sarma, Azadirachta indica leaf powder as a biosorbent for Ni(II) in aqueous medium, J. Hazard. Mater. 165 (2004) 271–278.

[332] S.S. Shukla, L.J. Yu, K.L. Dorris, A. Shukla, Removal of nickel from aqueous solutions by sawdust, J. Hazard. Mater. 121 (2005) 243–246.

[333] P. Yin, Q. Xu, R. Qu, G. Zhao, Removal of transition metal ions from aqueous solutions by adsorption onto a novel silica gel matrix composite adsorbent, J. Hazard. Mater. 169 (2009) 228–232.

[334] V. Padmavathy, Biosorption of nickel(II) ions by baker's yeast: kinetic, thermodynamic and desorption studies, Bioresour. Technol. 99 (2008) 3100–3109.

[335] M.E. Argun, S. Dursun, K. Gur, C. Ozdemir, M. Karatas, S. Dogan, Nickel adsorption on the modified pine tree materials, Environ. Technol. 26 (2005) 479–487.

[336] A. Ates, A. Yildiz, N. Yildiz, A. Calimli, Heavy metal removal from aqueous solution by Pseudevernia furfuracea (L.) Zopf, Ann. Chim. 97 (2007) 385–393.

[337] E. Malkoc, Ni(II) removal from aqueous solutions using cone biomass of Thuja orientalis, J. Hazard. Mater. 137 (2006) 899–908.

[338] H. Kalavathy, B. Karthik, L.R. Miranda, Removal and recovery of Ni(II) and Zn(II) from aqueous solution using activated carbon from Hevea brasiliensis: batch and column studies, Colloids. Surf. B Biointerfaces 78 (2010) 291–302.

[339] N. Boujelben, J. Bouzid, Z. Elouear, Adsorption of nickel and copper onto natural iron oxide-coated sand from aqueous solutions: study in single and binary systems, J. Hazard. Mater. 163 (2009) 376–382.

[340] X. Deng, Q.B. Li, Y.H. Lu, D.H. Sun, Y.L. Huang, Uptake of nickel from industrial wastewater by genetically engineered *Escherichia coli* JM109, Sheng. Wu. Gong. Cheng. Xue. Bao. 19 (2003) 343–348.

[341] M. Revathi, B. Kavitha, T. Vasudevan, Removal of nickel ions from industrial plating effluents using activated alumina as adsorbent, J. Environ. Sci. Eng. 47 (2005) 1–6.

[342] M. Iiyama, S. Oshima, H. Kokusen, M. Sekita, S. Tsurubou, Y. Komatsu, Adsorption of divalent transition metal ions with a chelating agent on octadecyl silica gel, Anal Sci. 20 (2004) 1463–1464.

[343] V.T. Vinod, R.B. Sashidhar, B. Sreedhar, Biosorption of nickel and total chromium from aqueous solution by gum kondagogu (Cochlospermum gossypium): a carbohydrate biopolymer, J. Hazard. Mater. 178 (2010) 851–860.

[344] M. Kobya, E. Demirbas, E. Senturk, M. Ince, Adsorption of heavy metal ions from aqueous solutions by activated carbon prepared from apricot stone, Bioresour. Technol. 96 (2005) 1518–1521.

[345] I. Ahmad, M.I. Ansari, F. Aqil, Biosorption of Ni, Cr and Cd by metal tolerant *Aspergillus niger* and Penicillium sp. using single and multi-metal solution, Indian J. Exp. Biol. 44 (2006) 73–76.

[346] N. Dizge, B. Keskinler, H. Barlas, Sorption of Ni(II) ions from aqueous solution by Lewatit cation-exchange resin, J. Hazard. Mater. 167 (2009) 915–926.

[347] Z. Chen, W. Ma, M. Han, Biosorption of nickel and copper onto treated alga (Undaria pinnatifida): application of isotherm and kinetic models, J. Hazard. Mater. 155 (2008) 327–333.

[348] K. Jayaram, M.N. Prasad, Removal of Pb(II) from aqueous solution by seed powder of Prosopis juliflora DC, J. Hazard. Mater. 169 (2009) 991–997.

[349] O.S. Lawal, A.R. Sanni, I.A. Ajayi, O.O. Rabiu, Equilibrium, thermodynamic and kinetic studies for the biosorption of aqueous lead(II) ions onto the seed husk of Calophyllum inophyllum, J. Hazard. Mater. 177 (2010) 829–835.

[350] N. Liu, D. Lin, H. Lu, Y. Xu, M. Wu, J. Luo, B. Xing, Sorption of lead from aqueous solutions by tea wastes, J. Environ. Qual. 38 (2009) 2260–2266.

[351] W.S. Ngah, S. Fatinathan, Pb(II) biosorption using chitosan and chitosan derivatives beads: equilibrium, ion exchange and mechanism studies, J. Environ. Sci. (China) 22 (2010) 338–346.

[352] E. Pehlivan, T. Altun, S. Cetin, M. Iqbal Bhanger, Lead sorption by waste biomass of hazelnut and almond shell, J. Hazard. Mater. 167 (2009) 1203–1208.

[353] Y. Zhang, Y. Li, X. Li, L. Yang, X. Bai, Z. Ye, L. Zhou, L. Wang, Selective removal for Pb^{2+} in aqueous environment by using novel macroreticular PVA beads, J. Hazard. Mater. 181 (2010) 898–907.

[354] H. Lalhruaitluanga, K. Jayaram, M.N. Prasad, K.K. Kumar, Lead(II) adsorption from aqueous solutions by raw and activated charcoals of Melocanna baccifera Roxburgh (bamboo) – a comparative study, J. Hazard. Mater. 175 (2010) 311–318.

[355] B. Southichak, K. Nakano, M. Nomura, N. Chiba, O. Nishimura, Marine macroalga Sargassum horneri as biosorbent for heavy metal removal: roles of calcium in ion exchange mechanism, Water Sci. Technol. 58 (2008) 697–704.

[356] M. Horsfall Jr., F. Ogban, E.E. Akporhonor, Biosorption of Pb(II) from aqueous solution by waste biomass of aerial roots of Rhizophora mangle (red mangrove), Chem. Biodivers. 2 (2005) 1246–1255.

[357] Y. Bulut, Z. Baysal, Removal of Pb(II) from wastewater using wheat bran, J. Environ. Manage. 78 (2006) 107–113.

[358] Q.W. Dai, F.Q. Dong, W. Zhang, Biosorption of lead ions on dried waste beer yeast and the analysis by FTIR, Guang. Pu. Xue. Yu. Guang. Pu. Fen. Xi. 29 (2009) 1788–1792.

[359] B. Sandrine, N. Ange, B.A. Didier, C. Eric, S. Patrick, Removal of aqueous lead ions by hydroxyapatites: equilibria and kinetic processes, J. Hazard. Mater. 139 (2007) 443–446.

[360] Y. Liu, X. Chang, Y. Guo, S. Meng, Biosorption and preconcentration of lead and cadmium on waste Chinese herb Pang Da Hai, J. Hazard. Mater. 135 (2006) 389–394.

[361] M. Amini, H. Younesi, N. Bahramifar, A.A. Lorestani, F. Ghorbani, A. Daneshi, M. Sharifzadeh, Application of response surface methodology for optimization of lead biosorption in an aqueous solution by *Aspergillus niger*, J. Hazard. Mater. 154 (2008) 694–702.

[362] M. Nadeem, A. Mahmood, S.A. Shahid, S.S. Shah, A.M. Khalid, G. McKay, Sorption of lead from aqueous solution by chemically modified carbon adsorbents, J. Hazard. Mater. 138 (2006) 604–613.

[363] X.M. Li, D.X. Liao, X.Q. Xu, Q. Yang, G.M. Zeng, W. Zheng, L. Guo, Kinetic studies for the biosorption of lead and copper ions by Penicillium simplicissimum immobilized within loofa sponge, J. Hazard. Mater. 159 (2–3) (2008) 610–615.

[364] F. Saiano, M. Ciofalo, S.O. Cacciola, S. Ramirez, Metal ion adsorption by Phomopsis sp. biomaterial in laboratory experiments and real wastewater treatments, Water Res. 39 (2005) 2273–2280.

[365] S.C. Wu, X.L. Peng, K.C. Cheung, S.L. Liu, M.H. Wong, Adsorption kinetics of Pb and Cd by two plant growth promoting rhizobacteria, Bioresour. Technol. 100 (2009) 4559–4563.

[366] Y. Bulut, Z. Tez, Removal of heavy metals from aqueous solution by sawdust adsorption, J. Environ. Sci. (China) 19 (2007) 160–166.

[367] J. Yu, M. Tong, X. Sun, B. Li, Enhanced and selective adsorption of Pb(II) and Cu(II) by EDTAD-modified biomass of baker's yeast, Bioresour. Technol. 99 (2008) 2588–2593.

[368] H. Benhima, M. Chiban, F. Sinan, P. Seta, M. Persin, Removal of lead and cadmium ions from aqueous solution by adsorption onto micro-particles of dry plants, Colloids Surf. B. Biointerfaces 61 (2008) 10–16.

[369] C. Chen, J. Wang, Removal of Pb(II), Ag(II), Cs(I) and Sr(II) from aqueous solution by brewery's waste biomass, J. Hazard. Mater. 151 (2008) 65–70.

[370] M. Martínez, N. Miralles, S. Hidalgo, N. Fiol, I. Villaescusa, J. Poch, Removal of lead(II) and cadmium(II) from aqueous solutions using grape stalk waste, J. Hazard. Mater. 133 (2006) 203–211.

[371] A.H. Hawari, C.N. Mulligan, Biosorption of lead(II), cadmium(II), copper(II) and nickel(II) by anaerobic granular biomass, Bioresour. Technol. 97 (2006) 692–700.

[372] M.R. Sangi, A. Shahmoradi, J. Zolgharnein, G.H. Azimi, M. Ghorbandoost, Removal and recovery of heavy metals from aqueous solution using Ulmus carpinifolia and Fraxinus excelsior tree leaves, J. Hazard. Mater. 155 (2008) 513–522.

[373] B.C. Qi, C. Aldrich, Biosorption of heavy metals from aqueous solutions with tobacco dust, Bioresour. Technol. 99 (2008) 5595–5601.

[374] G. Yan, T. Viraraghavan, Heavy-metal removal from aqueous solution by fungus Mucor rouxii, Water Res. 37 (2003) 4486–4496.

[375] K.N. Ghimire, K. Inoue, K. Ohto, T. Hayashida, Adsorption study of metal ions onto crosslinked seaweed Laminaria japonica, Bioresour. Technol. 99 (2008) 32–37.

[376] M. Khosravi, R. Rakhshaee, M.T. Ganji, Pre-treatment processes of Azolla filiculoides to remove Pb(II), Cd(II), Ni(II) and Zn(II) from aqueous solution in the batch and fixed-bed reactors, J. Hazard. Mater. 127 (2005) 228–237.

[377] M.M. Rao, D.K. Ramana, K. Seshaiah, M.C. Wang, S.W. Chien, Removal of some metal ions by activated carbon prepared from Phaseolus aureus hulls, J. Hazard. Mater. 166 (2009) 1006–1013.

[378] W.B. Lu, J.J. Shi, C.H. Wang, J.S. Chang, Biosorption of lead, copper and cadmium by an indigenous isolate Enterobacter sp. J1 possessing high heavy-metal resistance, J. Hazard. Mater. 134 (2006) 80–86.

[379] Saeed, A., Iqbal, M. and Akhtar, M.W. Removal and recovery of lead(II) from single and multimetal (Cd, Cu, Ni, Zn) solutions by crop milling waste (black gram husk), J. Hazard. Mater. 117 (2005) 65–73.

[380] O.D. Uluozlu, A. Sari, M. Tuzen, M. Soylak, Biosorption of Pb(II) and Cr(III) from aqueous solution by lichen (Parmelina tiliaceae) biomass, Bioresour. Technol. 99 (2008) 2972–2980.

[381] J.R. Rangel-Mendez, R. Monroy-Zepeda, E. Leyva-Ramos, P.E. Diaz-Flores, K. Shirai, Chitosan selectivity for removing cadmium(II), copper(II), and lead(II) from aqueous phase: pH and organic matter effect, J. Hazard. Mater. 162 (2009) 503–511.

[382] R. Zein, R. Suhaili, F. Earnestly, M.E. Indrawati, Removal of Pb(II), Cd(II) and Co(II) from aqueous solution using Garcinia mangostana, J. Hazard. Mater. 181 (2010) 52–56.

[383] W.Z. Zhou, W.W. Li, Y.Z. Zhang, B.Y. Gao, J. Wang, Biosorption of Pb(II) and Cu(II) by an exopolysaccharide from the deep-sea psychrophilic bacterium Pseudoalteromonas sp. SM9913, Huan. Jing. Ke. Xue. 30 (2009) 200–205.

[384] H. Salehizadeh, S.A. Shojaosadati, Removal of metal ions from aqueous solution by polysaccharide produced from Bacillus firmus, Water Res. 37 (2003) 4231–4235.

[385] L. Zhou, J. Liu, Z. Liu, Adsorption of platinum(IV) and palladium(II) from aqueous solution by thiourea-modified chitosan microspheres, J. Hazard. Mater. 172 (2009) 439–446.

[386] I. Smiciklas, S. Dimovic, M. Sljivic, I. Plecas, The batch study of Sr(II) sorption by bone char, J. Environ. Sci. Health A Tox. Hazard Subst. Environ. Eng. 43 (2008) 210–217.

[387] O.A. Zaporozhets, L.S. Ivanko, I.V. Marchenko, E.V. Orlichenko, V.V. Sukhan, Quercetin immobilized on silica gel as a solid phase reagent for tin(IV) determination by using the sorption-spectroscopic method, Talanta 55 (2001) 313–319.

[388] S. Xie, J. Yang, C. Chen, X. Zhang, Q. Wang, C. Zhang, Study on biosorption kinetics and thermodynamics of uranium by Citrobacter freudii, J. Environ. Radioact. 99 (2008) 126–133.

[389] K. Fujiwara, A. Rameshm, T. Maki, H. Hasegawa, K. Ueda, Adsorption of platinum (IV), palladium (II) and gold (III) from aqueous solutions onto L-lysine modified crosslinked chitosan resin, J. Hazard. Mater. 146 (2007) 39–50.

[390] G. Palma, J. Freer, J. Baeza, Removal of metal ions by modified Pinus radiata bark and tannins from water solutions, Water Res. 37 (2003) 4974–4980.

[391] Y.S. Al-Degs, M.I. El-Barghouthi, A.A. Issa, M.A. Khraisheh, G.M. Walker, Sorption of Zn(II), Pb(II) and Co(II) using natural sorbents: equilibrium and kinetic studies, Water Res. 40 (2006) 2645–2658.

[392] Y. Zhang, Y. Li, L.Q. Yang, X.J. Ma, L.Y. Wang, Z.F. Ye, Characterization and adsorption mechanism of Zn(II) removal by PVA/EDTA resin in polluted water, J. Hazard. Mater. 178 (2010) 1046–1054.

[393] E. Pehlivan, G. Arslan, Comparison of adsorption capacity of young brown coals and humic acids prepared from different coal mines in Anatolia, J. Hazard. Mater. 138 (2006) 401–408.

[394] S. Tunali, T. Akar, Zn(II) biosorption properties of Botrytis cinerea biomass, J. Hazard. Mater 131 (1–3) (2006) 137–145.

[395] S. Sirianuntapiboon, T. Hongsrisuwan, Removal of Zn(II) and Cu(II) by a sequencing batch reactor (SBR) system, Bioresour. Technol. 98 (2007) 808–818.

[396] H. Mohapatra, R. Gupta, Concurrent sorption of Zn(II), Cu(II) and Co(II) by Oscillatoria angustissima as a function of pH in binary and ternary metal solutions, Bioresour. Technol. 96 (2005) 1387–1398.

[397] A.H. Oren, A. Kaya, Factors affecting adsorption characteristics of Zn(II) on two natural zeolites, J. Hazard. Mater. 131 (2006) 59–65.

[398] P. King, K. Anuradha, S.B. Lahari, K.Y. Prasanna, V.S. Prasad, Biosorption of zinc from aqueous solution using Azadirachta indica bark: equilibrium and kinetic studies, J. Hazard. Mater. 152 (2008) 324–329.

[399] R.R. Sheha, Sorption behavior of Zn(II) ions on synthesized hydroxyapatites, J. Colloid Interface Sci. 310 (2007) 18–26.

[400] C. Chen, J.L. Wang, Cation K(I), Mg(II), Na(I), Ca(II) release in Zn(II) biosorption by *Saccharomyces cerevisiae*, Huan. Jing. Ke. Xue. 27 (2006) 2261–2267.

[401] F.V. Pereira, L.V. Gurgel, L.F. Gil, Removal of Zn(II) from aqueous single metal solutions and electroplating wastewater with wood sawdust and sugarcane bagasse modified with EDTA dianhydride (EDTAD), J. Hazard. Mater. 176 (2010) 856–863.

[402] E. Pehlivan, S. Cetin, B.H. Yanik, Equilibrium studies for the sorption of zinc and copper from aqueous solutions using sugar beet pulp and fly ash, J. Hazard. Mater. 135 (2006) 193–199.

3

Water Treatment for Organic Pollutants by Adsorption Technology

CHAPTER OUTLINE

Environmental Water, http://dx.doi.org/10.1016/B978-0-444-59399-3.00003-9

The detail of adsorption has already discussed in the previous chapter. Therefore, this chapter has directly described the application of adsorption technology for the removal of organic pollutants. The applications have been divided into two parts, i.e. adsorbents used prepared by organic and inorganic precursors. The direct application of adsorption technology for the removal of organic pollutants is discussed in the following subsections.

3.1 Organic Precursor Adsorbents

The various adsorbents have been prepared from different waste materials of organic origin. The preparation protocol of the adsorbents has not been discussed in detail as it is outside the domain of this chapter. The applications of various waste product in the preparation of adsorbents and water treatment are discussed herein.

3.2 Coconut Husk

Coconut shell is a well-known precursor for the production of high-quality granular activated carbons (ACs). About ~9% w/w global production of commercial AC is prepared from this material [1]. It has been considered as an inexpensive raw material due to its widespread distribution in the developing countries. Mortley et al. [2] prepared AC from coconut husks and shells and compared them with commercially available carbons. The precursors with higher activation energies yielded higher surface areas and developed microporosities on account of the lesser degree of lignocellulosic breakdown. Similarly, Laine et al. [3] developed carbons from Venezuelan coconut by chemical activation. The shells were impregnated with H_3PO_4 followed by one-step carbonization/activation at 450 °C. The surface area of the product was 1200 m^2/g. The chemically modified AC was used for the removal of rotenone from synthetic and real wastewaters by Dhaouadi et al. [4]. AC was treated with $(NH_4)_2S_2O_8$ and NH_3 and was observed that ammonia treated carbon has more removal capacity. Vieira et al. [5] used the coconut shell for the removal of textiles dyes of Blue Remazol R160 (BR 160), Rubi S2G (R S2G), Red Remazol 5R (RR 5), Violet Remazol 5R (VR 5), and Indanthrene Olive Green (IOG). The sorption capacity decreased in the order of BR 160 >VR 5 > RR 5 > R S2G > IOG. Hitchcock et al. [6] reported the adsorption of $ZnCl_2$ activated coconut coir. Adsorption of 4-nitrophenol was studied on the activated coir and the large amount of adsorbate is retained by these materials. Banerjee et al. [7] prepared AC by heating $ZnCl_2$ impregnated crushed coconut shells to 700 °C. The so obtained charred intermediate was activated in steam or air at 900 °C. The carbon so obtained was having surface area of 800 m^2/g with high adsorbing properties.

3.3 Fruits Waste

Fruit wastes are also important precursors for developing low-cost adsorbents. Olive stones and almond shells have been evaluated for the removal of some industrial dyes viz., Methylene Blue (MB), Orange II, Crystal Violet, and Victoria Blue from aqueous phase by Linares-Solano et al. [8]. Olive stones, almond shells, apricot and peach stones, palm fruit bunch, etc. are the prevailing raw materials for the preparation of AC. Nasser [9] prepared an adsorbent from palm fruit branch. The equilibrium and kinetic studies were carried out during the adsorption of basic yellow dye on fruit bunch adsorbent. The adsorbent was inexpensive and has a mono layer equilibrium capacity of 320 mg dye/g. Nassar et al. [10] also studied the ability of palm fruit bunch particles for the uptake of basic dye (BR 18) over a range of initial dye concentration with adsorbent particle sizes varying in the range 106–300 μm. The maximum adsorption capacity of palm bunch was 242 mg/g. A range of chemically activated Tamarind nut carbons were prepared by Srinivasan et al. [11] using a variety of treatments adapted from the work of Hassler [12]. Cardoso et al. [13] reported the removal of a textile dye Remazol black B from aqueous effluent by Araucaria angustifolia, a Brazilian pine fruit shell. Rodriguez-Reinso et al. [14] have optimized the activation procedures for AC production using a variety of chemical and physical techniques. The high-quality, microporus carbons have been prepared and characterized [14]. Langmuir equation was applied to evaluate the surface areas and it ranges between 90 and 1550 m^2/g. Caturla et al. [15] showed the efficacy of olive stone carbons for the uptake of nitro- and chloro-substituted phenols from aqueous solutions. The efficiency of the prepared material for the adsorption varied according to the degree of activation.

3.4 Bark and Other Tannin Materials

The bark and other tannin-rich materials are also good source for organic adsorbents. Bark is a solid waste of timber industry and it can be used as a possible adsorbent due to its high tannin content. The polyhydroxy polyphenol groups of tannin are found to be the active species in the adsorption process. The uptake of cations takes place by ion exchange by displacing the adjacent phenolic hydroxyl groups forming a chelate [16,17]. The main problem associated with tannin-containing materials is decoloration of the water due to soluble phenols. To overcome this problem, chemical pretreatment of bark has been tried by a number of workers. Edgehill and Lu [18] reported the potential of carbonized slash pine bark as a suitable substitute for AC. The bark was carbonized by slow heating in nitrogen atmosphere for 6.5 h at 672 °C. BET N_2 surface area, average micropore and mesopore diameter, and micropore volume of the prepared adsorbent were 332 m^2/g, 2.17 A°, and 0.125 cm^3/g, respectively. Portuguese pine bark showed suitable precursor for AC production by Guedes de Carvalho et al. [19]. Carbonization of ground bark was carried out at 600 °C for 1 h under a N_2 atmosphere and then activated between 800and 1000 °C by partial CO_2 gasification. Eucalyptus bark has been investigated for its potential use in the removal of reactive dyes (Remazol BB) [20]. It was observed that all factors studied had a significant effect on the adsorption process in the order of initial dye concentration > bark concentration > initial pH > sodium chloride concentration > temperature. Similar experiments were under taken with commercial AC and the results showed that adsorption capacity of bark is about half of the AC. Bras et al. [21] reported the ability of pine bark to remove organochlorine pesticides from aqueous solutions. The results showed that approximately 97% of heptachlor, aldrin, endrin, dieldrin, DDT, DDD, and DDE were removed from 1 to 10 μg/L solutions. Lindane could not be effectively adsorbed (38% yield of removal). The authors also compared pine bark with AC and suggested that for the compounds with similar physicochemical properties, pine bark can be employed as an alternative to carbon. Li et al. [22] reported the removal efficiency of polycyclic aromatic hydrocarbons (PAHs).

Alves et al. [23] reported formaldehyde pretreatment to diminish the bleeding of colored compounds without appreciably affecting capacity. Randall et al. [16] reported a comparative study of bark adsorption to that of peanut skins, walnut expeller meal, and coconut husks and found their adsorption capacities to be comparable. Orhan and Buyukgungor [24] observed the adsorption capacities of the tannin having products obtained from nut and walnut shell, waste tea, and coffee and compared it with that of AC. They demonstrated that the capacities of the tannin-containing products are only slightly less than that of AC. Mckay et al. [25] examined the utility of five inexpensive adsorbents, i.e. bark, cotton, rice, hair, husk, and coal for the adsorption of two cationic dyes, i.e. safranine and MB. The monolayer adsorption capacities for safranine were higher than MB on all the adsorbents studied except that on coal. Zhou et al. [26] removed the neutral red dye (cationic dye) from the aqueous solution using cottonseed hull, a low-cost widely available agricultural waste in China. Ibrahim et al. [27] modified barley straw with NaOH and a cationic surfactant hexadecylpyridinium chloride monohydrate (CPC) to remove two anionic dyes: acid blue (AB 40) and reactive blue 4 (RB 4) from aqueous solution. CPC was observed to strongly attach to straw surface Removal percentage of AB 40 and RB 4 was quite good. The adsorption of these dyes on modified straw was favorable at high acidic condition but desorption process was found relatively less efficient.

3.5 Waste Tyres

Waste tyres are important for adsorbent and these waste tyres have high carbon content. Thus, controlled pyrolysis has been employed to convert the butyl rubber and carbon black content obtained from these waste tyres to semi-active carbon. Stripped tyres rubber has been used by Lucchesi and Maschio [28] to obtain carbon using moving bed reactor at 400–700 °C and a variety of atmospheres. Surface area of the finished carbon was found up to 320 m^2/g. This product was used for the removal of Orange II and Acid Black 24 dyes from aqueous phase. For vapor phase applications, a highly active carbon of surface area 1260 m^2/g was prepared by Ogassawara et al. [29] using a wet thermal process (water/steam 700–900 °C). Paprowicz [30] prepared powdered AC (PAC) from waste tyre rubber. The surface area of the product was quite low (193 m^2/g) but the sorbent exhibited a considerable affinity for the removal of aqueous phase phenol during subsequent batch adsorption trials. Streat et al. [31] carried out a comparative study of the sorption of phenol and *p*-chlorophenol from water using novel samples prepared by carbonization and subsequent activation of straw and used rubber tyres and conventional ACs based on coal, coconut shell, and wood. The used tyre samples were carbonized in a tube furnace I in a stream of N_2 at 3 °C/min to 800 or 900 °C. Further the product was activated using a stream of moist nitrogen at fixed temperature by varying the gas flow rate and the reaction time. The sorption of phenol and *p*-chlorophenol is almost identical to that of conventional ACs.

3.6 Saw Dust and Other Wood Materials

Some other tannin-rich agriculture byproducts have been tested as low-cost adsorbents. Hardwood (Beech) sawdust with 13% moisture content was used as an adsorbent by Asfour et al. [32] for the basic dye astrazone blue FRR 69. The results showed that the capacity of hardwood for adsorption of basic dye reduced with temperature, i.e. from 82.2 mg dye/g of wood at 25 °C to 105.7 mg dye/g of wood at 80 °C. The data showed that the relative cost of hardwood sawdust was 8.4% of AC. Therefore, there was no need to recover the sawdust and the exhausted wood can be used as a fuel. Gupta et al. [33] studied the adsorption–desorption of indigocarmine using deoiled mustard. Desorption was significantly enhanced up to 85% when using 30% glycerol as eluting agent. Mckay and Poots [34] used saw dust waste for the removal of telon and astrazone blue dyes. It was found that for telon blue removal, the cost using wood was only 1.5% to that of carbon, and for astrazone blue removal, the cost was only 8.2% to that of carbon.

3.7 Rice Husk

The rice husks or hulls are significant high-volume agricultural wastes for the production of low-cost adsorbents. Youseff et al. [35] used chemical activation of rice husk carbon (RHC) by ($ZnCl_2$) and steam activation. The chemically activated materials reported (600 °C, limited air) have the capacities of 0.24–0.72 mM for MB which depends on the initial $ZnCl_2$ impregnation ratio than steam-ACs (900 °C, 12–43% burn off). But it has poor affinity for this dye (0.05–0.28 mM). Manique et al. [36] studied the purification of biodiesel from the waste frying oil (WFO) by rice husk ash. The impurity of WFO ranges from 1% to 5% in the biodiesel. The best removal of WFO was found when its concentration was 4% in the biodiesel. The adsorption was attributed to the presence of high concentration of silica as well as the presence of meso and macro pores in the composition of WFO. Tannin and Gurgey [37] studied water purification by RHC and MB adsorption on RHC was fruitfully achieved. Nawar and Doma [38] determined the adsorptive capacities of peat moss and raw rice hulls for the removal of two industrial textile dyes, Sandocryl Orange B-3RLE and Lanasyn Black BRL ABK and compared these with those of a granular commercial carbon. The maximum adsorption was observed with the peat moss (95% w/w, 50 mg/L initial concentration, 2-h contact time) followed by rice hulls (90% w/w) and the least with commercial carbon (87% w/w). The relative costs of peat moss and raw rice hulls for the removal of textile dyes were be 0.52% and 1.19% in comparison to carbon for Sandocryl Orange and 1.4% and 0.3% for Lanazyn black, respectively.

3.8 Miscellaneous Agricultural Wastes

Cellulosic and lignocellulosic wastes have long been found as good raw materials for the preparation of AC. An agricultural waste in India has been utilized to prepare light-weight charcoal chips by Banerjee and Mathew [39]. Simitzis and Sfyrakis [40,41] have examined carbon obtained by the activation of phenol–formaldehyde resin mixed in a 20/80 blend with both olive stones and cottonseeds. The products showed similar adsorption capacities as demonstrated by commercial carbon for 3-nitrophenol, 4-nitrophenol, and MB. Incorporation of the resin gives added support to the AC. Tahn et al. [42] utilized decaying leaves of a common urban-planted plane (*Plantannus orientalis*) and found them capable for the removal of phenol and buffering excessive acid base from aqueous solutions. The removal processes were found slightly dependent on both pH and ionic strength. Han et al. [43] studied the removal of MB by lotus leaf from the aqueous solution. The adsorption on the lotus leaf was due to hydroxyl and carbonyl group at the surface of leaf. The adsorption capacity of the lotus leaf was 221.7 mg/g. Sun and Xu [44] reported efficacy of sunflower and stalks as an adsorbents for MB, basic red 9 (BR), congo red, and direct blue 71 from aqueous solutions. The adsorption rates of two basic dyestuffs were much higher than that of the direct dyes.

Perineau et al. [45] reported lignocellulosic component of wool carbonizing waste for the removal of dyes from textile wastewater. The uptake of dyes was selective and can be explained in terms of ionic chemisorptive interactions between dye molecule and charged ammonical sorbent substrate. Lopez-Gonzalez et al. [46,47] reported the preparation of AC from Spanish olive wood obtained from the pruning and felling of olive groves. The product was capable of adsorbing considerable amounts of 4-nitrophenol from aqueous solutions. A series of carbon have been prepared by Martinez-Sanchez et al. [48] using chromium-tanned leather waste. This precursor was high in ash (Cr_2O_3) content, i.e. 22–69 weight wise percent. It was carbonized and activated at 825 °C in a CO_2 atmosphere to produce a carbon with a maximum surface area of 108 m^2/g. Renouprez and Avom [49] prepared AC from palm tree cobs, a waste from the palm-oil industry in Cameroon. The ACs obtained possessed surface areas ranging 23–1078 m^2/g^{-1}.

3.9 Fertilizer Industry Wastes

The waste slurry of fertilizer plants has been converted into inexpensive adsorbents material by Srivastava et al. [50]. The waste slurry was first treated with hydrogen peroxide and then heated to 200 °C till the evolution of black soot stopped. The heated product was cooled and activated in the presence of air at 450 °C for 1 h, which yielded an adsorbent of 630 m^2/g surface area. This product exhibited good adsorption potential for substituted phenols [51,52]. The studied phenols were 2,4,6-trinitrophenol, 4-nitrophenol, 4-chlorophenol, phenol, 1,3-dihydroxybenzene, and 1,3,5-trihydroxybenzene [51]. The results indicated that 2,4,6-trinitrophenol had the highest affinity for the adsorbent while 1,3,5-trihydroxybenzene had the least. The process was economically feasible as the product was inexpensive than the commercially available AC. The competitive effect on the uptake of these substituted phenols viz. on the carbonaceous adsorbent has been studied using the Langmuir and Snoeyink modified model [52]. For phenolic compounds studied, the rate was higher for 2,4,6-trinitrophenol than for the other substituted derivatives.

Their relative rates decreased in the order of 4-nitrophenol > 4-chlorophenol > phenol > 1,3-dihydroxybenzene. Srivastava et al. [53] also reported the adsorption of dinitrophenol (DNP) on this adsorbent. The removal of DNP was 100% w/w at low concentrations while it was 80% w/w at higher concentration at optimum pH (4.0). Both batch and column studies were performed. Some feasibility experiments have been performed with an aim to recover DNP and chemical regeneration of the spent column. It was reported that 60 mL of 5% NaOH was sufficient for almost complete desorption of DNP (desorbed 55.2 mg). After regeneration with 1 M HNO_3 the sorption capacity of the column was almost the same as that of virgin adsorbent material.

3.10 Seafood Wastes

Generally, chitin is found in the exoskeletons of crabs and other arthropods and the cell wall of some fungi [54,55]. A large amount of chitin is produced from crab meat canning industry. More than 40,000 tones/annum of chitin is available from the fisheries of crustacean [56]. Chitosan or glucosamine is a deacetylated derivative of chitin and can be chemically prepared from chitin. Chitosan is found in the cell wall of some fungi like Mucorales stains and can be used as an effective adsorbent. About 1–4 million pounds of chitosan could be produced at a cost of $1.00–2.00 per pound [54]. Liu et al. [57] synthesized a series of chitosan-g-poly (acrylic acid)/vermiculite hydrogel composites and used as adsorbents for the removal of MB from aqueous solution. It was found that 10 wt % vermiculite in the chitosan-g-poly (acrylic acid) polymeric network, hydrogel composite thus obtained showed the highest adsorption capacity for MB, and could be regarded as a possible adsorbent for cationic dye removal in aqueous medium. Akar and Divriklioglu [58] reported decolorization of Reactive Red 2 by modified macro fungus in wastewater. The chitosan beads were used for the removal of perfluorooctane sulfonate from aqueous solution [59]. Biosorption efficiency of the adsorbent increased with a small amount of fungal biomass after the modification process and the loading capacity was 141.53 mg dye/g. Another material studied for its pollutants adsorbing nature was chitin. Chitin appears to be more economically attractive as an adsorbent since it was abundant in nature.

3.11 Petroleum Industry Wastes

Large quantities of carbonaceous waste residues generated in the petroleum refineries attracted many scientists. These wastes have been utilized as a precursor for AC. Kogyo and Kaisha [60] reported a method demonstrating the production of pitch beads (a mixture of petroleum pitch and polyaromatics). The initial oxidation of pitch beads was carried out at 400 °C and the resultant product was activated at 800 °C to obtain the adsorbent of optimum quality. Fasoli and Genon [61] used a refinery waste for the preparation of a granular carbon. The precursor was spent sand filter which was used for the removal of insoluble oil residues from a refinery wastewater. The waste was carbonized in a rotary furnace and then activated under steam atmosphere. Reuse of the carbon for the same application was also suggested by the authors. Ali and Saleem [62] reported heavy vacuum bottoms obtained from an Iraqi crude oil for the production of a carbonaceous adsorbent.

3.12 Sea Weed and Algae

It is well-known that some seaweed, i.e. brown algae have significant ion exchange properties in their polysaccharide contents. The seaweeds have shown high sorption capacities. Holan et al. [63] observed that the biomass has a tendency to disintegrate and swell, which can be effectively used for column operations. Modifications of seaweeds by cross-linking increase the stability and mechanical properties. The cross-linking retarded the swelling so that the material would be amenable to column adsorption. Rathinam et al. [64] reported the Brown seaweed biomass as a precursor of AC; the AC was then characterized and used as an adsorbent for the removal of phenol from aqueous solution. The best conditions for the removal of phenol by seaweed were determined at pH 3.0 with 150 mg/L, 10 g/L, 4 h, 50 °C, 75 strokes/min as initial concentration of phenol, adsorbent dosage, time, temperature, and an agitation speed, respectively. The maximum uptake of 98.31% of phenol was found under these experimental conditions. Gupta et al. [65] used aeromonas formicans bacteria for the removal of COD from a pulp mill wastewater. The authors reported the removal of COD up to 71–78%.

3.13 Peat Moss

The peat moss is considered as cheap and widely available adsorbent, with adsorption capacities for a variety of pollutants. The peat is mined and processed for fuel in Ireland and Northern Europe and is used as a soil conditioner in the United States. An average cost for a standard pack (501) of sphagnum peat moss is $7.75 [66]. Basically, the peat is a complex soil material with decomposed organic matter. Lignin and cellulose are major constituents of peat. These contents, especially lignin, contain polar functional groups that can be involved in chemical bonding. Because of these properties, peat moss tends to have a high cation capacity and can be effectively used as an adsorbent for the removal of a variety of pollutants [67]. Peat is highly porous (95%) and possess a large specific surface area ($>200\ m^2/g$) [68]. There are different types of peat available with varying chemical and physical properties. Mckay et al. [69] reported a three-step model for the adsorption of astrazone blue (basic blue (BB 69)) on peat.

3.14 Inorganic Precursor Adsorbents

3.14.1 Fly Ash

The estimated global production of fly ash is 290 metric tons per annum [70] and there is no set methodology to use it and a large amount is dumped as land filler which is still 10%

w/w of the total mass available. Fly ash is an inorganic residue and its chemical constituents comprise of 40–50% w/w in silica (SiO_2) 20–35% w/w in alumina (Al_2O_3) and 5–12% w/w iron oxide (Fe_2O_3). Flyash also contains 12–30% w/w carbon and unburnt residue [71]. The removal of dyes and phenols from industrial wastewaters using fly ash is reported [72], Paprowicz 1990 [30]). Kumar et al. [73] reported the removal efficiencies of phenol, cresol, and phenol:cresol (1:1) using fly ash as an adsorbent. Chrome red dye adsorption was investigated by Gupta et al. [74] using locally available fly ash. They reported that 99% removal from 5 mg/L solution within 2-h time. The same group [75] demonstrated complete removal of chrome red dye from 10 mg/L initial concentration by a mixture of fly ash and coal in a 1:1 blend. Sun et al. [76] reported the adsorption of anionic dyes (Reactive Red 23, RB 171, Acid Black 1, and AB 193) on fly ash and reported that the adsorption was pH dependent with the maximum adsorption capacity at the initial solution pH 7.5–8.5 for reactive dyes and 5–6 for acid dyes. The optimum time to reach the equilibrium was 60 min for both dyes with an increase in the initial dye concentration enhanced the adsorption capacity, but failed to increase the dye removal efficiency.

Albanis et al. [77] reported the adsorption and desorption of four selected chlorophenols and two herbicides on fly ash. The mass balance showed that the adsorbed amounts of chlorophenols in mixtures of soil with 30% flyash were up to 36.5% for 2,4-dichlorophenol, 55.8% for 2,4,6-trichlorophenol, 68.7% for 2,3,5,6 tetrachlorophenol, and 84.9% for pentachlorophenol and the respective adsorbed amounts of herbicides in mixtures with 20% flyash were up to 37.5% for alachlor and 43.2% for metachlor. Contrarily, the amounts of desorption decreased as the increase in flyash content and the increase in the number of chlorine groups in the phenolic ring. Alemany et al. [78] reported the possible use of flyash for the removal of phenol from water. Aksu and Yener [79] used dried activated sludge and fly ash for removing of *o*-chlorophenol and *p*-chlorophenol. The pollutant binding capacity of the adsorbent/biosorbent was found to depend on function of the substituted group, pH, and mono-chlorinated phenol concentration. The pH values were determined as 1.0 and the equilibrium uptake increased with increasing initial mono-chlorinated phenol concentration up to 500 mg/dm^3 for all the mono-chlorinated phenol-sorbent systems. Freundlich, Langmuir, and Redlich–Peterson models were used to the equilibrium data and mono chlorinated phenol-sorbent system was observed. The results showed Redlich–Peterson model as the best model for all the mono-chlorinated phenol-sorbent systems. Kapoor and Viraraghavan [80] studied the properties of fly ash as extremely variable and the adsorption capacity varied with the lime content. Daniels et al. [81] reported the feasibility of utilizing alkaline flyash to control acid mine drainage. The results showed that this technique has potential for industrial scale application. Banerjee et al. [82] described the adsorption kinetics of *o*-xylene by flyash. Mahadeva et al. [83] reported coal flyash as a nonconventional, low-cost adsorbent for the removal of phenol. Experiments were conducted under varying conditions, viz. adsorbent dose, pH, initial concentration, and contact time. At equilibrium, 95–96% adsorption on coal flyash adsorbent was achieved.

3.14.2 Blast Furnace Slag

The steel plants produce large volume of granular blast furnace slag as a by-products; causing a serious disposal problem. However, it is being used as filler or in the production of slag cement. The blast furnace slag adsorbent is inexpensive as the waste is available. It has been converted into an effective and economical adsorbent for the removal of some toxic pollutants from water as well as wastewater. Gupta et al. [74,75] reported the removal of malachite green, a dye, broadly used in a textile industry by its adsorption onto activated blast furnace slag. Blast furnace slag was obtained from Tata Iron and Steel Company, Limited, Jamshedpur (India). The waste was washed with water and dried at 200 °C. The heated product was cooled and activated in air at 600 °C for 1 h. The surface area of the finished product was 107 m^2/g and the studies revealed that the developed activated slag could be employed as adsorbent for the removal of dyes. The removal of malachite green was almost 99.9% at lower concentrations. The column capacity was found to be higher than batch capacity.

3.14.3 Metal Oxides and Hydroxides

Solisio et al. [84] reported the adsorption of two different oils and preliminary tests carried out with an excess of the material showed that the insoluble oil was adsorbed. The material activation with HCl improved the adsorption of soluble oil organic fraction, while a little increase in the removal was for the insoluble oil. The results showed that the products of dolomite calcination could be replaced. The conventional adsorbing materials in the removal of organic pollutants, with particular concern to exhausted soluble oils, which could not be recycled. Therefore, the operational cost reduced of this treatment. Bernar et al. [85] studied the adsorption of various organic micropollutants on pyrolusite (MnO_2). Poulton et al. [86] used iron oxide/hydroxides for the removal of dissolved hydrogen sulfide from seawater under flow-through conditions. Ferrihydrite was stabilized by precipitation onto zeolite pellets, and rates of sulfide removal were determined under laboratory conditions at pH 8.5. Sulfide removal kinetics was dependent on the initial sulfide concentration, substrate mass, and flow rate. The results from laboratory experiments compared favorably with sulfide removal kinetics determined in a series of experiments performed online in a recirculating mariculture production system. The presence of phosphate reduced 50% in sulfide removal for substrate removed from an online system. Ferrihydrite-coated zeolite worked as an efficient, cheap adsorbent for sulfide removal, which was suited to small-scale aqueous flow-through systems. The reaction of iron oxides/hydroxides with dissolved sulfide was accompanied by a distinct color change due to the formation of black FeS.

Asuha et al. [87] reported the removal of methyl orange (MO) on mesoporous TiO_2. Mesoporous TiO_2 was prepared by a hydrothermal method using cetyltrimethyl ammonium bromide as a structure-directing agent and the formed mesoporous TiO_2 (5.2 nm as an average pore size and a surface area of 161.2 m^2/g). The maximum adsorption capacity and adsorbed percentage of the mesoporous were be 454.5 mg/g and 87%.

3.14.4 Sugar Industry Wastes

Sugar industries produce a large amounts of waste products such as suspended solids, organic matter, press mud, bagasse pith, and bagasse fly ash. These wastes especially of bagasse fly ash do not find any use as such and causes a disposal problem. Currently, this is being used as filler in building materials. Bagasse pith is a significant waste product form the sugar cane industry and generally regarded as having very little economic value. The pith was not subjected to any pretreatment prior to use. A standard agitated tank configuration was used for the batch contact studies. The maximum saturation capacities of bagasse pith for the two dyes were 21.7 mg dye/g for AB 25 and 22.9 mg dye/g for acid red 114 (AR 114). In another study, Mckay et al. [88] studied the adsorption of four dyes onto bagasse pith. Bagasse fly ash and other waste product have been used as adsorbents for the removal of many contaminants from the water. Srivastava et al. [89,90] prepared adsorbent from bagasse flyash for the removal of substituted phenols. The uptake of trinitrophenol on the bagasse flyash was compared with other commercially available adsorbents. It was found that the developed bagasse flyash adsorbent was inexpensive. The removal of DNP has also been observed by Gupta et al. [91] and the cost estimation of bagasse flyash showed about Rs. 60/ton. By considering the cost of transport, chemicals, electrical energy used in the process the finished product it became about Rs. 400/ton. Therefore, the adsorbent may be a good replacement of commercially available carbon due to its comparable efficiency and low cost. The results showed that 86% of DNP could be removed using 10 g/L of activated bagasse flyash. Mckay et al. [92] studied the adsorption of two acid dyes viz. AR 114 and AB 25 onto bagasse pith have been removed.

Gupta et al. [93] used bagasse fly ash for the removal of pesticides, i.e. lindane and malathion from wastewater. The authors reported the maximum uptake 97–98% at pH 6, 5 g/L dose. The results were analyzed with the Langmuir and Freundlich models. The thermodynamic parameters were also calculated and the adsorption reported was exothermic in nature. Furthermore, Gupta et al. [94] reported the removal of phenol and *p*-nitrophenol from wastewater using activated bagasse fly ash. The batch studies showed 95% of phenol and 99.9% *p*-nitrophenol removed by using 10 g/dm^3 of adsorbent at pH 4.0. The adsorption capacity K_F was less for the phenol–bagasse flyash system than that for the *p*-nitrophenol–bagasse flyash system. Langmuir constant Q° decreased with increase in temperature thereby indicating the process to be exothermic in nature. The desorption was obtained with a number of eluting agents and almost complete desorption (~98%) of phenol could be achieved with only 8% NaOH. Gupta et al. [94,95] have removed successfully basic dyes, i.e. rhodamine B and Methylene Blue and pesticides, i.e. DDD and DDE from wastewater and aqueous solutions using a sugar industry waste—bagasse fly ash. Khattri and Singh [96] observed the removal of malachite green, Methylene Blue, rhodamine B, and crystal violet using sugarcane dust.

3.14.5 Clays

Fuller's earth is being used industrially for its decolorizing properties globally as it is an inexpensive and naturally abundant material. There are many kinds of clays but montmorillonite may be expected to have the highest sorptive capacity in comparison to other. El-Guendi [97] reported the adsorption of BB 69 and BR 22 from aqueous solutions on natural clay. El-Guendi et al. [98] reported the adsorption of cationic dyestuffs (BB 69 and BR 22) from aqueous solutions by natural and activated clays. The maximum adsorption capacities (q_{max}) of BB 69 and BR 22 on to natural clay were 390 and 365 mg/g, respectively. The natural clay was treated with H_2O_2 and HCl to enhance its adsorption capacity for dyestuffs. The efficiency of activation with H_2O_2 was greater by 23% and 13% in the case of BB 69 and BR 22, respectively, in comparison to natural clay. The efficiency of activation with HCl was greater by 30% and 16% in the case of BB 69 and BR 22, respectively, relative to natural clay. Gupta and Shukla [99] reported adsorption techniques for the treatment of carpet effluents containing mainly Metomega Chrome Orange GL using various low-cost materials, i.e. coal, flyash, china clay, and wollastonite. Low initial dye concentration, temperature, and pH favored the removal process. Mckay et al. [100] described the adsorption capacity of Fuller's earth for basic and AB to be 220 and 120 mg/g, respectively. Thus, they proposed the adsorbent for the treatment of textile effluents. The adsorption capabilities of clay were due to negative charge on the structure of fine grain silicate minerals. This negative charge was neutralized by the adsorption of positively charged cations such as dyes etc. Besides, the clays showed a large surface area up to 800 m^2/g, which contributes to its high-adsorption capacity [101].

Lee et al. [102] modified montmorillonite and reported the removal of 2-chlorophenol, 3-cynophenol and 4-nitrophenol from water. This resulted in the change in surface property of montmorillonite from hydrophilic to organophilic. Polubesova et al. [103] also reported interactions of phenolic acids (ferulic, *p*-coumaric, syringic, and vanillic) with Fe(III)-montmorillonite. Adsorption of the various phenolic acids on Fe(III) enriched montmorillonite was accompanied by their oxidative transformation and formation of Fe(II). Similar findings were reported by Kim et al. [104]. Sorption of phenol and 2-, 3-, and 4-chlorophenol from water by tetramethylammonium (TMA^+)-smectite and tetramethylphosphonium (TMP^+)-smectite was reported by Lawrance et al. [105]. TMP-smectite seems to be a good sorbent. Besides, TMP-smectite possessed selective sorption within the group of chlorinated phenols examined. Phenol and 4-chlorophenols were effectively sorbed by TMP-smectite. Contrarily, 2- and 3-chlorophenols could not be sorbed. The adsorption–desorption of 2,4,6- trichlorophenol (TCP) on calcium montmorillonite (pure clay), prepared montmorillonite humic complexes, and natural soils were investigated by Sabbah and Rebhum [106]. Daneis et al. [107] determined the adsorption properties of alumina pillared montmorillonite (PMt) and mesoporous alumina aluminum phosphates for the removal of 2,4- dichlorophenol, 2,4,6-trichlorophenol, and pentachlorophenol. Malusis et al. [108] reported the adsorption of phenol on soil-bentonite (SB) vertical barriers amended with AC. SB backfills was amended with 0–10 wt % granular AC (GAC) or PAC. The AC amended backfills exhibited enhanced phenol adsorption than unamended backfill due to hydrophobic partitioning to the AC. Adsorption capacity increased with increasing AC content but does not change for the GAC and PAC. Gil et al. [109] studied the removal of orange II and MB by two-pillared clay formed by the intercalation of two pillared of aluminum (Al-PILC) and zirconium (Zr-PILC). For orange II, both the pillared clays have same adsorption capacity while for the MB the adsorption capacity of Zr-PILC was higher than that of Al-PILC. It was found that the to pore size distribution had affected the efficiency and selectivity of adsorption of two dyes. Clay could be modified to enhance its efficiency for the removal of pollutants from water and wastewaters. Cadena et al. [101] modified bentonite by replacing the natural exchangeable cations present in the clay by the organophilic cations, TMA^+ ion. Results revealed that the adsorption capacity of natural bentonite were improved by heat treatment while the acid treatment decreased the same. The adsorption of four types of amino acids (lysine, glycine, glutamic acid, and alanine) on sodium and calcium bentonite has been reported [110].

Boyd et al. [111] reported the mechanisms of the adsorption of substituted nitrobenzenes onto smectite clays. Cheknane et al. [112] reported the adsorption of two basic dyes, i.e. CI Basic Yellow 28 (BY 28) and CI Basic Green 4 (BG 4) in single and binary solute systems using two classes of inorganic–organic pillared clay granules as sorbents prepared by granulation from an Al cetyltrimethylammonium bromide intercalated clay powder. Adsorption rate data indicated that BY 28 adsorbs more rapidly than BG 4 but the adsorption of BG 4 was greater than that of BY 28 in single solute and binary dye systems. Errais et al. [113] reported the textile dye removal by Fouchana Tunisian clay in leaching textile and dye industry. The adsorbent contains 60% smectites, 30% kaolinite, and 10% illite, having the cation-exchange capacity of about 50 meq/100 g of clay. Ho et al. [114] reported the kinetics of BR 18 and AB 9 sorption onto activated clay. Activation energy of sorption was also determined based on the pseudosecond-order rate constants. Wu et al. [115] reported the adsorption of phenol by both inorganic and organic PMts. The adsorbing capacity of modified clays depended not only surface area but mainly on micropore structure and surface components. The modified PMt with surfactant was found to improve adsorbing capacity greatly. The PMt can be recycled, and it is a potential substance for adsorption of environmental pollutants.

3.14.6 Sediment and Soil

The sediment sand and soil have also been used for the removal of organic pollutants from water. The soils have been applied to remove the environmental pollutants by adsorption. Morillo et al. [116] reported the adsorption of glyphosate (GPS) adsorption on three soils of different characteristics. They observed that the interaction of this pesticide with the soils was not related to their CEC and clay minerals content, but to the content of iron and aluminum amorphous oxides and organic matter. The presence of copper in treatment solutions increased GPS adsorption, due to reasons as: GPS coordinates strongly to Cu, and Cu–GPS complexes formed seem to have higher ability to be adsorbed on the soil than free GPS; GPS adsorption as Cu was previously adsorbed, acting as a bridge between the soil and GPS. The adsorption and desorption of PAHs on different soils was investigated by Walter et al. [117] via static and dynamic methods. On the basis of a system including the four phases of soil, water, oil adsorbed, and oil in emulsion, a model for the description of the adsorption behavior in the presence of oil was reported. The distribution coefficients were determined from the octanol/water coefficient of the aromatic compounds and the organic carbon content of the soils. For the oil-contaminated system, the modeling of the adsorption behavior enabled a more detailed interpretation of the experimental observations as well as the calculation of the sorption behavior for the PAHs from characteristic parameters of the components involved. Kibe et al. [118] studied the adsorption of pesticides on the soil. The adsorption equilibriums of the five herbicides, i.e. esprocarb, mefenacet, pretilachlor, simetryn, and thiobencarb, on five kinds of paddy soil in Japan were investigated. It was found that equilibrium values were better expressed by the Freundlich equation for concentration levels for the paddy fields and that the values for the adsorption coefficient (n) varied from 1.0 to 1.6. The values of the coefficient (k) were in the range of 29–420 $mg^{1-1/n}\ L^{1/n}$/kg-dry and poorly related to solubilities in water or to octanol–water partition coefficients of the herbicides. The adsorption equilibriums calculated from the values of adsorption constant K_d by the values of K_{OC} in the literature were observed to be different from the experimental equilibriums. Jalil et al. [119] reported the removal of MO from aqueous solution using calcined Lapindo volcanic mud (LVM) as an adsorbent. The maximum adsorption capacity of LVM was 333.3 mg/g. Gao et al. [120] reported the removal of phenolic compounds, i.e. 2,4-dichlorophenol, 2, 4-DNP, and 2, 4-dimethyphenol on the river sediment.

3.14.7 Red Mud

The red mud (RM) is a waste material formed during the production of alumina when the bauxite ore is subjected to caustic leaching. A typical bayer process plant produces a large quantity of RM (1–2 tons RM per ton of alumina produced [121]) and the toxicity and colloidal nature of RM particles create a serious environmental problem. There have been many proposals for RM management, i.e. in the manufacture of RM bricks, as filler in asphalt road construction, as iron ore, and as a source of various minerals [122]. A salt-free preparation of the RM had specific surface area of 58 m^2/g. The BET surface areas for AC, bone char, activated bauxite, chi-alumina, and gamma alumina are 802, 117, 155, 312, and 178 m^2/g, respectively. Bhatnagar et al. [123] reviewed the removal efficiency of the RM for different aquatic pollutants. RM is very good absorbent for the removal of different types of dyes, phenols, and other organic pollutants. These experiments indicated that after pretreatment, RM was suitable for the treatment of wastewaters. Lambert and Graham [124] investigated the amenability of nonspecific natural organic compounds for removal by adsorption on to AC, activated alumina, activated bauxite, and bone char adsorbents. Among them activated bauxite has a heterogeneous array of varying pore radii and was as effective as both aluminas for the removal of nonspecific organic compounds and humic acid compounds. Further, water treated RM was found suitable for maximum adsorption among adsorbents developed from RM. The ability of RM for the adsorption of congo red dye from aqueous solutions was determined by Namasivayam and Arasi [125]. Lopez et al. [126] reported the feasibility of RM for wastewater treatment. RM was obtained from Alumina-Aluminio of San Ciprian (Lugo, Spain).

3.14.8 Zeolites and Ore Minerals

Zeolites silicate minerals found in nature and can also be synthesized at commercial level. They clinoptilolite is the most abundant of more than 40 natural zeolite species [127], which occurs in abundance throughout the western United States, easily available and inexpensive. The ion-exchange capabilities determine the adsorption properties of zeolites. Phenol adsorption on natural phillipsite was only effective on NH_4^+ exchanged phillipsite [128]. The bioremedition of pyridine and quinoline by biozeolite is carried out. The ammonium ion is often generated during the biodegradation of pyridine and quinoline because of the transformation of N from pyridine and quinoline. Bai et al. [129] used a specific biozeolite composed of mixed bacteria for degrading pyridine and quinoline. The modified zeolite was used for biodegradation and adsorption in two types of wastewater, i.e. sterile synthetic and coking wastewater. Both pyridine and quinoline degraded simultaneously by the mixed bacteria. The ammonium ion transformed from pyridine and quinoline could be removed by the modified zeolite. In this way, biozeolite may be used for the removal in the treatment of wastewater containing pyridine and/or quinoline. Dong et al. [130] prepared zeolite from coal fly ash (ZFA), modified with hexadecyltrimethylammonium (HDTMA) and used for the adsorption of bisphenol A (BPA) from water. It was observed that HDTMA formed bilayer micelles on external surfaces of zeolites. Thus, an enhanced adsorption capacity of ZFA had no affinity for BPA. Adsorption of BPA by ZFA was improved slightly in the presence of NaCl and at a low temperature.

Ore minerals were found suitable for the removal of a variety of organic pollutants by adsorption. Bouyarmane et al. [131] used natural phosphate rock and two synthetic mesoporous hydroxyapatites for the removal of pyridine and phenol from aqueous solution. Both natural and synthetic apatites indicated similar pyridine sorption capacities whereas phenol loading was proportional with their respective specific surface area. This was due to the strong interaction between pyridine and apatite surface that hinders further interparticular diffusion. Bouyarmane et al. [131] reported that, in spite of low specific surface area, natural phosphate rock can be used as an efficient sorbent material for various pollutants for better efficiency and lower processing costs than some ACs. Some of the adsorptions of organic pollutants are summarized in Table 3-1.

Table 3-1 Adsorption of Organic Pollutants on Different Adsorbents

Adsorbents	Contaminants	Concentration Range	Adsorption Capacities (mg/g)	% Adsorption	Contact Time	Dose	pH	References
Bagasse pith	AR 114, AB 25 BB 69, BR 22	200 mg/dm^3	22.9, 21.7, 157.4, 76.6	–	5 days	–	–	[92]
Sunflower stalk	MB, BR 9, congo red, direct blue	100–2000, 100–2000, 50–1000, 50–1000 mg/L	205, 317, 37.78, 26.84	80%	5 days	0.20 g/50 ml	–	[37]
Hardwood sawdust	Astrazone	200 mg/dm^3	82.2–105.7	–	3 h	–	–	[32]
Carbonaceous material	2,4,6-Tri nitrophenol, 4- nitrophenol, 4-chlorophenol, 1,3-dihydroxy benzene	1×10^{-4} to 1×10^{-3} M	8.0, 5.32, 4.57, 3.38	–	24 h	–	2–10	[51]
Carbonaceous material	2,4-DNP	10^{-4} to 10^{-3} M	8.0×10^{-4}	100–80%	10 h	0.1 g/10 ml	4.0	[53]
Fly ash	Phenol, cresol, phenol: cresol	50 ppm	–	40%, 45%, and 45%	–	–	–	[73]
Fly ash	Chrome red dye	5 mg/L	–	99%	2 h	–	–	[74]
Fly ash	Xylene	50–150 mg/L	–	72–84%	35 min	30 g/L	–	[82]
Bagasse fly ash	Rhodamine B, MB	1.0×10^{-5} to 1.0×10^{-4}, 1.0×10^{-3} to 1.0×10^{-2} M	–	60–100%	8 h	10.0, 4.0 g/L	2.0, 8.0	[94]
Bagasse pith	AR 114, AB 25	100 mg/dm^3	–	–	48 h	3.4 g	–	[88]
Bagasse fly ash	Phenol, *p*-nitrophenol	10^{-4} to 10^{-3} M	–	95%, 100%	24 h	10 g/dm^3	4.0	[91]
Bagasse fly ash	2,4-DNP	10^{-4} to 10^{-3} M	–	86%	6–7 h	10 g/L	2–4	[90]
Sugarcane dust	Malachite green, crystal violet, MB, rhodamine B	12.0 mg/L	3.999, 3.798, 3.745, 3.240	83.32%, 79.13%, 78.02%, 67.50%	30 min	0.5 g/200 ml	7.5	[96]

Peat particles	Astrazone blue	200 mg/dm^3	0.43000–9100	–	48 h	0.2 g/50 × 10^{-3} dm^3	4.0–4.3	[69]
Kaolnite	Metomega Chrome Orange	10 mg/L	0.6506	84.23%	120 min	1.0 g/50 ml	4.5	[98]
Wollastonite	Metomega Chrome Orange	10 mg/L	0.6957	73.04%	150 min	1.0 g/50 ml	4.5	[98]
Fly ash	Chrome Orange	10 mg/L	0.7428	92.25%	110 min	1.0 g/50 ml	4.5	[99]
Slag	Malachite green	1×10^{-4} to 1×10^{-3} M	74.16	~100–70%	6–8 h	1 g/L	6–10	[132]
Carbonaceous material	Malachite green	1×10^{-4} to 1×10^{-3} M	75.08	~100–66%	6–8 h	1 g/L	2–6	[132]
Animal bone, incinerator ash	*p*-Nitrophenol	200 mg/L	111, 116	–	–	0.01, 0.001 g	2.77–8.80	[133]
Rhizopus arrhizus and *Calluna vulgaris*	Iron(III)-cyanide	1996.2, 387 mg/L	612.2, 387	–	–	–	13.0	[134]
Charcoal	Triton	0–5 mM		90%, 95%, 95%	2 h	1 g/100 ml	–	[135]
Maize cob	AB 25, AR 114	0.05 dm^3	41.4, 47.7	–	5 days		2.0	[136]
Spent oil shale	Phenol	10–200 ppm	1.0	–	–	2–40 g/50 ml	3–11	[137]
HDTMA-montmorillonite	Phenol, 2-nitrophenol, 3-nitrophenol, 4 nitrophenol	–	140.1, 43.03, 128.0, 81.52	–	2 h	0.5 g	–	[138]
Blue magnetite	Acridine orange, MB, neutral red, safranino, thionine, congo red	1 mg/ml	105.40, 78.08, 55.19, 75.54, 106.40, 2.805	–	2 h	–	–	[139]
Silicone resin	Isoprothiolane	–	–	97%	–	–	–	[140]
Buffing dust carbon	Acid brown dye	50–125 mg/L	2.84–6.24	–	60 min	1 g	3.5	[141]
40% TMA/45% HDTMA-montmorillonite	2-Chlorophenol, 3-cynophenol, 4-nitrophenol	50–2500 mg/L	45.1, 34.4, 46.8	–	–	0.501 g		[102]
Sitka spruce wood	Telon blue, astrazone blue	100, 200 mg/dm^3	9.3–15.6, 71.8–80.1	–	120 min	–	–	[34]

Conclusions

The applications of low-cost adsorbent for the removal of organic pollutants were thus found to be remarkable. The removal efficiencies of these inexpensive adsorbents are quite good and ranged up to 99.9%. Besides, the free availability of precursors for these adsorbents make them ideal candidates for water treatment. Their applications are important for developing and under developed countries. It can be concluded from the critical evaluation of this chapter that the discussed low-cost adsorbents have wide range of applications in removal of different organic contaminants. There are many publications on this issue but the technique is not fully developed, especially at pilot and industrial scales. Therefore, there is a great need of the removal of organic pollutants from water by using low-cost adsorbents at pilot and industrial scales.

References

[1] R.P. Bansal, J.P. Donnet, F. Stoeckli, Active Carbon, Marcel Dekker, New York, 1988.

[2] Q. Mortley, W.A. Mellowes, S. Thomas, Activated carbons from materials of varying morphological structure, Thermochim. Acta. 129 (1988) 173–186.

[3] J. Laine, A. Calafat, M. Labady, Preparation and characterization of activated carbons from coconut shell impregnated with phosphoric acid, Carbon 27 (1989) 191–195.

[4] A. Dhaouadi, L. Monser, N. Adhoum, Removal of rotenone insecticide by adsorption onto chemically modified activated carbons, J. Hazard. Mater. 181 (2010) 692–699.

[5] A.P. Vieira, S.A.A. Santana, C.W.B. Bezerra, H.A.S. Silva, J.A.P. Chaves, J.C.P. Melo, et al., Removal of textile dyes from aqueous solution by babassu coconut epicarp (Orbignya speciosa), Chem. Eng. J. 173 (2011) 334–340.

[6] S.J. Hitchcock, B. McEnaney, S.J. Watling, Fibrous active carbons from coir, J. Chem. Technol. Biotechnol. 33A (1983) 157–163.

[7] S.K. Banerjee, S. Majumdar, A.C. Dutta, A.K. Roy, S.C. Banerjee, D.K. Banerjee, Active carbon from coconut shell, Ind. J. Technol. 14 (1976) 45–49.

[8] A. Linares-Solano, J. de D. Lopez-Gonzalez, M. Molina-Sabio, F. Rodriguez-Reinoso, Active carbons from almond shells as adsorbents in gas and liquid phases, J. Chem. Technol. Biotechnol. 30 (1980) 65–72.

[9] M.M. Nasser, The kinetics of basic dye removal using palm fruit bunch, Adsorp. Sci. Technol. 14 (1997) 609–617.

[10] M.M. Nasser, M.F. Hamoda, G.H. Radwan, Utilization of palm-fruit bunch particles for the adsorption of dye stuff wastes, Adsorp. Sci. Technol. 13 (1996) 1–6.

[11] K. Srinivasan, P.B.S. Rao, A. Ramadevi, Studies on characteristics of carbon obtained from Tamarind nuts, Ind. J. Environ. Health 30 (1988) 303–305.

[12] J.W. Hassler, Activated Carbon, Leonard Hill, London, 1967.

[13] N.F. Cardoso, R.B. Pinto, E.C. Lima, T. Calvete, C.V. Amavisca, B. Royer, et al., Removal of remazol black B textile dye from aqueous solution by adsorption, Desalination 269 (2011) 92–103.

[14] F. Rodriguez-Reinso, J. de D. Lopez-Gonzalez, C. Berenguer, Activated carbons from almond shell: I-preparation and characterization by nitrogen adsorption, Carbon 20 (1982) 513–518.

[15] F. Caturala, J.M. Martin-Martinez, M. Molina-Sabio, F. Rodriguez-Reinso, R. Torregrosa, Adsorption of substituted phenols on activated carbons, J. Colloid Interface Sci. 124 (1988) 528–534.

[16] J.M. Randall, R.L. Bermann, V. Garrett, A.C. Waiss Jr., Use of bark to remove heavy metal ions from waste solutions, For. Prod. J. 24 (1974) 80–84.

[17] G. Vazquez, G. Antorrena, J. Gonzalez, M.D. Doval, Adsorption of heavy metal ions by chemically modified Pinus Pinaster bark, Bioresource Technol. 48 (1994) 251–255.

[18] R.U. Edgehill, G.Q. Lu, Adsorption characteristics of carbonized bark for phenol and pentachlorophenol, J. Chem. Technol. Biotechnol. 71 (1998) 27–37.

[19] R.A. Guedes de Carvalho, C.G. Gonzalez-Beca, M.N. Sampaio, O. Neves, M.C.S. Pereira, A. Macedo, Use of pine bark for preparation of activated carbon and as a soil conditioner, Agric. Wastes 9 (1984) 231–238.

[20] L.C. Morais, O.M. Freitas, E.P. Goncalves, L.T. Vasconcelos, C.G.G. Beca, Reactive dyes removal from wastewaters by adsorption on eucalyptus bark: variables that define the process, Water Res. 33 (1999) 979–988.

[21] I.P. Bras, L. Santos, A. Alves, Organochlorine pesticides removal by pinus bark sorption, Environ. Sci. Technol. 33 (1999) 631–634.

[22] Y. Li, B. Chen, L. Zhu, Enhanced sorption of polycyclic aromatic hydrocarbons from aqueous solution by modified pine bark, Bioresour. Technol. 101 (2010) 7307–7313.

[23] M.M. Alves, C.G. Gonzalez Beca, R. Guedes de Carvalho, J.M. Castanheria, M.C.S. Pereira, L.A.T. Vasconcelos, Chromium removal in tannery wastewaters polishing by Pinus sylverstris bark, Water Res. 27 (1993) 1333–1338.

[24] Y. Orhan, H. Buyukgungor, The removal of heavy metals by using agricultural wastes, Water Sci. Technol. 28 (1993) 247–255.

[25] G. McKay, J.F. Porter, G.R. Prasad, The removal of dye colors from aqueous solutions by adsorption on low-cost material, Water Air Soil Pollut. 114 (1999) 423–438.

[26] Q. Zhou, W. Gong, C. Xie, D. Yang, X. Ling, X. Yuan, et al., Removal of Neutral Red from aqueous solution by adsorption on spent cottonseed hull substrate, J. Hazard. Mater. 185 (2010) 502–506.

[27] S. Ibrahim, I. Fatimah, H.M. Ang, S. Wang, Adsorption of anionic dyes in aqueous solution using chemically modified barley straw, Water Sci. Technol. 62 (2010) 1177–1182.

[28] A. Lucchesi, G. Maschio, Semi-active carbon and aromatics produced from the pyrolysis of scrap tyres, Conserv. Recyc. 6 (1983) 85–90.

[29] S. Ogassawara, S.M. Kuroda, N. Wakao, Preparation of activated carbon by thermal decomposition of used automotive tyres, Ind. Eng. Chem. Res. 26 (1987) 2552–2556.

[30] J.T. Paprowicz, Activated carbons for phenols removal from wastewaters, Environ. Technol. 11 (1990) 71–81.

[31] M. Streat, J.W. Patrick, M.J. Camporro-Perez, Sorption of phenol and p-chlorophenol from water using conventional and novel activated carbons, Water Res. 29 (1995) 467–472.

[32] H.M. Asfour, O.A. Fadali, M.M. Nasser, M.S. EI-Geundi, Equilibrium studies on adsorption of basic dyes on hardwood, J. Chem. Technol. Biotechnol. 35A (1985) 21–27.

[33] V.K. Gupta, R. Jain, S. Malathi, A. Nayak, Adsorption-desorption studies of indigocarmine from industrial effluents by using deoiled mustard and its comparison with charcoal, J. Colloid Interface Sci. 348 (2010) 628–633.

[34] G. Mckay, V.J.P. Poots, Kinetics and diffusion processes in color removal from effluent using wood as an adsorbent, J. Chem. Technol. Biotechnol. 30 (1980) 279–292.

[35] A.M. Youseff, M.R. Mostafa, E.M. Dorgham, Surface properties and decolorizing power of carbons from rice husks, Afinidad 425 (1990) 41–44.

[36] M.C. Manique, C.S. Faccini, B. Onorevoli, E.V. Benvenutti, E.B. Caramão, Rice husk ash as an adsorbent for purifying biodiesel from waste frying oil, Fuel 92 (2012) 56–61.

[37] S. Tannin, I. Gurgey, Bleaching of cotton seed and sunflower oils by active carbons from rice hulls, Chim. Acta Turica. 16 (1988) 209–219.

[38] S.S. Nawar, H.S. Doma, Removal of dyes from effluents using low-cost agricultural by-products, Sci. Total Environ. 79 (1989) 271–279.

[39] S.K. Banerjee, M.D. Mathew, Carbonisation of jute stick, and agricultural waste, Agric. Waste 15 (1985) 225–229.

[40] J. Simitzis, J. Sfyrakis, Adsorption of phenols on carbonaceous adsorbents produced from phenol-formaldehyde resin mixed with olive stones, Angew. Makromol. Chem. 163 (1988) 47–61.

[41] J. Simitzis, J. Sfyrakis, Production of carbonaceous adsorbents by using novolac-resin and cotton seed, J. Appl. Polymer Sci. 36 (1988) 1769–1779.

[42] T.T. Than, Y.X. Wang, W.B. Li, Sorption of phenol from aqueous solutions and acid-base neutralization by decaying of plane tree, Toxicol. Environ. Chem. 56 (1996) 161–169.

[43] X. Han, W. Wang, X. Ma, Adsorption characteristics of methylene blue onto low cost biomass material lotus leaf, Chem. Eng. J. 171 (2011) 1–8.

[44] G. Sun, X. Xu, Sunflower stalk as adsorbents for color removal from textile wastewater, Ind. Eng. Chem. Res. 36 (1997) 808–812.

[45] F. Perineau, J. Molinier, A. Gaset, Adsorption of ionic dyes on charred plant material, J. Chem. Technol. Biotechnol. 32 (1982) 749–758.

[46] J. de D. Lopez-Gonzalez, C. Valenzuela-Calahorro, A. Navarette-Guijosa, V. Gomez-Serrano, Preparation of active carbons from olive wood, Adv. Sci. Technol. 2 (1985) 253–262.

[47] J. de D. Lopez-Gonzalez, C. Valenzuela-Calahorro, A. Navarette-Guijosa, V. Gomez-Serrano, Adsorption of p-nitrophenol by active carbons prepared from olive wood, Ann. Quim. Ser. B 84 (1988) 47–51.

[48] M.A. Martin-Sanchez, Evaluation of Microporosity in Activated Carbons with a High Ash (Cr2O3) Content, in IUPAC Symposium on the Characterization of Porous Solids (COPS-II), Preprints, Alicante, May 69, IUPAC, Universidad de Alicante, Spain, 90, 1990.

[49] A. Renouprez, J. Avom, Characterisation of active carbons from palm tree fibres using nitrogen adsorption and small angle X-ray scattering, in: K.K. Unger, J. Rouquerol, K.S.W. Singh (Eds.), Characterisation of Porous Solids, Elsevier Science Publishers, The Netherlands, 1988.

[50] S.K. Srivastava, R. Tyagi, N. Pant, Adsorption of heavy metal ions on carbonaceous material developed from the waste slurry generated in local fertilizer plants, Water Res. 23 (1989) 1161–1165.

[51] S.K. Srivastava, R. Tyagi, Removal of substituted phenols by activated carbon (developed from fertilizer waste) and some thermodynamic aspects, Fresenius Environ. Bull. 3 (1994) 12–17.

[52] S.K. Srivastava, R. Tyagi, Competitive adsorption of substituted phenols by activated carbon developed from the fertilizer waste slurry, Water Res. 29 (1995) 483–488.

[53] S.K. Srivastava, R. Tyagi, N. Pal, D. Mohan, Process development for removal of substituted phenol by carbonaceous adsorbent obtained from fertilizer waste, J. Environ. Eng. 123 (1997) 842–851.

[54] R.C.W. Berkeley, Chitin, chitosan and their degradative enzymes, in: R.C.W. Berkeley, C.W. Gooday, D.C. Elwood (Eds.), Microbial Polysaccharides, Academic Press, New York, 1979.

[55] G.L. Roorer, T.Y. Hsien, J.D. Way, Synthesis of porous-magnetic chitosan beads for removal of cadmium ions from wastewater, Ind. Eng. Chem. Res. 32 (1993) 2170–2178.

[56] E. Onsoyen, O. Skaugrud, Metal recovery using chitosan, J. Chem. Technol. Biotechnol. 49 (1990) 395–404.

[57] Y. Liu, Y. Zheng, A. Wang, Enhanced adsorption of Methylene Blue from aqueous solution by chitosan-g-poly (acrylic acid)/vermiculite hydrogel composites, J. Environ. Sci. 22 (2010) 486–493.

[58] T. Akar, M. Divriklioglu, Biosorption applications of modified fungal biomass for decolorization of Reactive Red 2 contaminated solutions: batch and dynamic flow mode studies, Bioresour. Technol. 101 (2010) 7271–7277.

[59] Q. Zhang, S. Deng, G. Yu, J. Huang, Removal of perfluorooctane sulfonate from aqueous solution by crosslinked chitosan beads: sorption kinetics and uptake mechanism, Bioresour. Technol. 102 (2011) 2265–2271.

[60] K.K. Kogyo, K. Kaisha, Method and apparatus for activating infusibilized pitch beads, Eur. Patent 317 (1988) 217.

[61] U. Fasoli, G. Genon, Activated carbon by pyrolysis of organic sludges, Water Res. 10 (1976) 545–547.

[62] L.H. Ali, F. Saleem, Conversion of heavy crude oil vacuum bottoms to activated carbons, J. Petrol Res. 7 (1988) 49–62.

[63] Z.R. Holan, B. Volesky, I. Prasetyo, Biosorption of cadmium by biomass of marine algae, Biotechnol. Bioeng. 41 (1993) 819–825.

[64] A. Rathinam, J.R. Rao, B.U. Nair, Adsorption of phenol onto activated carbon from seaweed: determination of the optimal experimental parameters using factorial design, J. Taiwan Inst. Chem. Eng. 42 (2011) 952–956.

[65] V.K. Gupta, A.K. Minocha, N. Jain, Batch and continuous studies on treatment of pulp mill wastewater by Aeromonas formicans, J. Chem. Technol. Biotechnol. 76 (2001) 547–552.

[66] Y.S. Ho, D.A.J. Wase, C.F. Forster, The adsorption of divalent copper ions from aqueous solution by sphagnum moss peat, Trans. Inst. Chem. Eng. 72B (1994) 185–194.

[67] Mineral Resources Institute Technical Report Series, Use of Alabama Peat as an Adsorbent for Heavy Metals, MRI Technical Report Series. T.R. No. 12 (1985).

[68] J.K. McLelland, C.A. Rock, Pretreating landfill leachate with peat to remove metals, Water Air Soil Pollut. 37 (1988) 203–215.

[69] G. McKay, S.J. Allen, I.P. McConvey, M.S. Otterburn, Transport processes in the sorption of colored ions by peat particles, J. Colloid Interface Sci. 80 (1981) 323–339.

[70] R.K. Dhir, Pulverized Fuel Ash in Cement Replacement Materials, Concrete, Technology and Design, in: R.N. Swamy (Ed.), vol. 3, Surrey University Press, UK, 1986.

[71] G. Ferraiolo, M. Zilli, A. Converti, Flyash utilization, J. Chem. Technol. Biotechnol. 47 (1990) 281–305.

[72] S.A. Boyd, Adsorption of substituted phenols by soil, Soil Sci. 134 (1982) 337–343.

[73] S. Kumar, S.N. Upadhyay, Y.D. Upadhyay, Removal of phenol by adsorption of flyash, J. Chem. Technol. Biotechnol. 37 (1987) 281–290.

[74] G.S. Gupta, G. Prasad, K.K. Panday, V.N. Singh, Removal of chrome dye from aqueous solutions by flyash, Water Air Soil Pollut. 37 (1988) 13–24.

[75] G.S. Gupta, G. Prasad, V.N. Singh, Removal of chrome dye from aqueous solutions by mixed adsorbents: flyash and coal, Water Res. 24 (1990) 45–50.

[76] D. Sun, X. Zhang, Y. Wu, X. Liu, Adsorption of anionic dyes from aqueous solution on fly ash, J. Hazard. Mater. 181 (2010) 335–342.

[77] T.A. Albanis, T.G. Danis, M.G. Kourgia, Adsorption-desorption studies of selected chlorophenols and herbicides and metal release in soil mixture with flyash, Environ. Technol. 19 (1998) 25–34.

[78] L.J. Alemany, M.C. Jimenez, M.A. Larrubia, F. Blasco, F. Delgado, Removal of phenol from aqueous solution by adsorption on to coal fly ash, Adsorp. Sci. Technol. 13 (1996) 527–536.

[79] Z. Aksu, J. Yener, A comparative adsorption/biosorption study of mono-chlorinated phenols onto various sorbents, Waste Manage. 21 (2001) 695–702.

[80] A. Kapoor, T. Viraraghavan, Treatment of metal industrial wastewater by flyash and cement fixation, J. Environ. Eng. 122 (1996) 243–258.

[81] W.L. Daniels, B. Stewart, M. Jackson, Utilization of flyash to prevent acid mine drainage from coal reuse, in: Proceedings of the 10th International Ash Use Symposium, (1993).

[82] K. Banerjee, P.N. Cheremisinoff, S.L. Cheng, Adsorption kinetics of o-xylene by flyash, Water Res. 31 (1997) 249–261.

[83] S.M. Mahedeva, I.D. Mall, B. Prasad, I.M. Mishra, Removal of phenol by adsorption on coal flyash and activated carbon, Pollut. Res. 16 (1997) 170–175.

[84] C. Solisio, A. Lodi, A. Converti, M.D. Borghi, Removal of exhausted oils by adsorption on mixed Ca and Mg oxides, Water Res. 36 (2002) 899–904.

[85] S. Bernard, P. Chazal, M. Mazet, Removal of organic compounds by adsorption on pyrolusite (b-MnO2), Water Res. 31 (1997) 1216–1222.

[86] S.W. Poulton, M.D. Krom, J.V. Rijn, R. Raiswell, The use of hydrous iron (III) oxides for the removal of hydrogen sulphide in aqueous systems, Water Res. 36 (2002) 825–834.

[87] S. Asuha, X.G. Zhou, S. Zhao, Adsorption of methyl orange and Cr(VI) on mesoporous TiO2 prepared by hydrothermal method, J. Hazard. Mater. 181 (2010) 204–210.

[88] G. Mckay, M. El-Geundi, M.M. Nasser, Equilibrium studies for the adsorption of dyes on bagasse pith, Adsorp. Sci. Technol. 15 (1997b) 251–270.

[89] S.K. Srivastava, V.K. Gupta, N. Johri, D. Mohan, Removal of 2,4,6,-trinitrophenol using bagasse fly ash a sugar industry waste material, Ind. J. Chem. Technol. 2 (1995a) 333–336.

[90] S.K. Srivastava, V.K. Gupta, I.S. Yadav, D. Mohan, Removal of 2,4, dinitrophenol using bagasse fly ash—a sugar industry waste material, Fresenius Environ. Bull. 4 (1995b) 550–557.

[91] V.K. Gupta, S. Sharma, I.S. Yadav, D. Mohan, Utilization of bagasse fly ash generated in the sugar industry for the removal and recovery of phenol and para-nitrophenol from wastewater, J. Chem. Technol. Biotechnol. 72 (1998) 180–186.

[92] G. McKay, M. El-Geundi, M.M. Nasser, Adsorption model for the removal of acid dyes from effluent by bagasse pith using a simplified isotherm, Adsorp. Sci. Technol. 15 (1997a) 737–752.

[93] V.K. Gupta, C.K. Jain, I. Ali, S. Chandra, S. Agarwal, Removal of lindane and malathion from wastewater using bagasse fly ash—a sugar industry waste, Water Res. 36 (2002) 2483–2490.

[94] V.K. Gupta, D. Mohan, M. Sharma, S. Sharma, Removal of basic dyes (rhodamine-B and methylene blue) from aqueous solutions using bagasse fly ash, Sep. Sci. Technol. 35 (2000) 2097–2113.

[95] V.K. Gupta, I. Ali, Removal of DDD and DDE from wastewater using bagasse fly ash—a sugar industry waste, Water Res. 35 (2001) 33–40.

[96] S.D. Khattri, M.K. Singh, Colour removal from dye wastewater using sugar cane dust as an adsorbent, Adsorp. Sci. Technol. 17 (1999) 269–282.

[97] M.S. El-Guendi, Adsorption kinetics of cationic dyestuffs on to natural clay, Adsorp. Sci. Technol. 13 (1996) 295–303.

[98] M.S. El-Guendi, H.M. Ismail, K.M.E. Attyia, Activated clay as an adsorbent for cationic dyestuffs, Adsorp. Sci. Technol. 12 (1995) 109–117.

[99] G.S. Gupta, S.P. Shukla, An inexpensive adsorption technique for the treatment of carpet effluents by low cost materials, Adsorp. Sci. Technol. 13 (1996) 15–26.

[100] G. Mckay, M.S. Otterburn, J.A. Aja, Fuller's earth and fired clay as adsorbents for dye stuffs, Water Air Soil Pollut. 24 (1985) 307–322.

[101] F. Cadena, R. Rizvi, R.W. Peters, Feasibility Studies for the Removal of Heavy Metals from Solution Using Tailored Bentonite, in Hazardous and Industrial Wastes, Proceedings of the Twenty-Second Mid-Atlantic Industrial Waste Conference, Drexel University (1990) 77.

[102] J.H. Lee, D.I. Song, Y.W. Jeon, Adsorption of organic phenols onto dual organic cation montmorillonite from water, Sep. Sci. Technol. 32 (1997) 1975–1992.

[103] T. Polubesova, S. Eldad, B. Chefetz, Adsorption and oxidative transformation of phenolic acids by Fe(III)-montmorillonite, Environ. Sci. Technol. 44 (2010) 4203–4209.

[104] Y.S. Kim, D.I. Song, Y.W. Jeon, S.J. Choi, Adsorption of organic phenols onto hexadecyltrimethyl ammonium treated montmorillonite, Sep. Sci. Technol. 31 (1996) 2815–2830.

[105] M.A.M. Lawrence, R.K. Kukkadapu, S.A. Boyd, Adsorption of phenol and chlorinated phenols from aqueous solution by tetra methyl ammonium and tetramethyl phosphonium exchanged montmorillonite, Appl. Clay Sci. 13 (1998) 13–20.

[106] I. Sabbah, M. Rebham, Adsorption-desorption of trichlorophenol in water soil systems, Water Environ. Res. 69 (1997) 1032–1038.

[107] T.G. Daneis, T.A. Albanis, D.E. Petrakis, P.J. Pomonis, Removal of chlorinated phenols from aqueous solutions by adsorption on alumina pillared clays and mesoporous alumina aluminum phosphate, Water Res. 32 (1998) 295–302.

[108] M.A. Malusis, J.E. Maneval, E.J. Barben, C.D. Shackelford, E.R. Daniels, Influence of adsorption on phenol transport through soil-bentonite vertical barriers amended with activated carbon, J. Contam. Hydrol. 116 (2010) 58–72.

[109] A. Gil, F.C.C. Assis, S. Albeniz, S.A. Korili, Removal of dyes from wastewaters by adsorption on pillarad clays, Chem. Eng. J. 168 (2011) 1032–1040.

[110] M. Fushan, W. Shiwer, Q. Yongning, W. Xiao, Adsorption of amino acid by bentonite, ion exchange and adsorption, Lizi Jiaohuan Yu Xifu. 13 (1997) 54–59.

[111] A. Boyd, G. Sheng, B.J. Teppen, C.T. Johnston, Mechanisms for the adsorption of substituted nitrobenzenes by smectite clays, Env. Sci. Technol. 35 (2001) 4227–4234.

[112] B. Cheknane, M. Baudu, J.P. Basly, O. Bouras, Adsorption of basic dyes in single and mixture systems on granular inorganic-organic pillared clays, Environ. Technol. 7 (2010) 815–822.

[113] E. Errais, J. Duplay, F. Darragi, Textile dye removal by natural clay—case study of Fouchana Tunisian clay, Environ. Technol. 31 (2010) 373–380.

[114] Y.S. Ho, C.C. Chiang, Y.C. Hsu, The kinetics of Basic Red 18 and Acid Blue 9 sorption onto activated clay, Sepn Sci. Technol. 36 (2001) 2473–2488.

[115] P.X. Wu, Z.W. Liao, H.F. Zhang, J.G. Guo, Adsorption of phenol on inorganic-organic pillared montmorillonite in polluted water, Env. Int. 26 (2001) 401–407.

[116] E. Morillo, T. Undabeytia, C. Maqueda, A. Ramos, Glyphosate adsorption on soils of different characteristics: influence of copper addition, Chemosphere 40 (2000) 103–107.

[117] T. Walter, H.J. Ederer, C. Först, L. Stieglitz, Sorption of selected polycyclic aromatic hydrocarbons on soils in oil-contaminated systems, Chemosphere 41 (2000) 387–397.

[118] K. Kibe, M. Takahashi, T. Kameya, K. Urano, Adsorption equilibriums of principal herbicides on paddy soils in Japan, Sci. Total Environ. 263 (2000) 115–125.

[119] A.A. Jalil, S. Triwahyono, S.H. Adam, N.D. Rahim, M.A. Aziz, N.H. Hairom, et al., Adsorption of methyl orange from aqueous solution onto calcined Lapindo volcanic mud, J. Hazard. Mater. 181 (2010) 755–762.

[120] P. Gao, Y. Feng, Z. Zhang, J. Liu, N. Ren, Comparison of competitive and synergetic adsorption of three phenolic compounds on river sediment, Environ. Pollut. 159 (2011) 2876–2881.

[121] G. Akay, B. Keskinler, A. Cakici, U. Danis, Phosphate removal from water by red mud using crossflow microfiltration, Water Res. 32 (1998) 717–726.

[122] Y.L. Loretta, Properties of red mud tailings produced under varying process conditions, J. Environ. Eng. 124 (1998) 254–265.

[123] A. Bhatnagar, V.J.P. Vilar, C.M.S. Botelho, R.A.R. Boaventura, A review of the use of red mud as adsorbent for the removal of toxic pollutants from water and wastewater, Environ. Technol. 32 (2011) 231–249.

[124] S.D. Lambert, N.J.D. Graham, Removal of non-specific dissolved organic matter from upland potable water supplies-I: aadsorption, Water Res. 29 (1995) 2421–2433.

[125] C. Namasivayam, D.J.S.E. Arasi, Removal of Congo red from wastewater by adsorption onto waste red mud, Chemosphere 34 (1997) 401–417.

[126] E. Lopez, B. Soto, M. Arias, A. Nunez, D. Rubinosand, M.T. Barral, Adsorbent properties of red mud and its use for wastewater treatment, Water Res. 32 (1998) 1314–1322.

[127] D.W. Ming, J.B. Dixon, Quantitative determination of clinoptilolite in soils by a cation-exchange capacity method, Clays Clay Miner. 35 (1987) 463–468.

[128] J.E. Garcia, M.M. Gonzalez, J.S. Notario, Phenol adsorption on natural phillipsite, Reactive Polymers 21 (1993) 171–176.

[129] Y. Bai, Q. Sun, R. Xing, D. Wen, X. Tang, Removal of pyridine and quinoline by bio-zeolite composed of mixed degrading bacteria and modified zeolite, J. Hazard. Mater. 181 (2010) 916–922.

[130] Y. Dong, D. Wu, X. Chen, Y. Lin, Adsorption of bisphenol A from water by surfactant-modified zeolite, J. Colloid Interface Sci. 348 (2010) 585–590.

[131] H. Bouyarmane, S. El Asri, A. Rami, C. Roux, M.A. Mahly, A. Saoiabi, et al., Pyridine and phenol removal using natural and synthetic apatites as low cost sorbents: influence of porosity and surface interactions, J. Hazard. Mater. 181 (2010) 736–741.

[132] V.K. Gupta, S.K. Srivastava, D. Mohan, Equilibrium uptake, sorption dynamics, process optimization and column operations for the removal and recovery of malachite green from wastewater using activated carbon and activated slag, Ind. Eng. Chem. Res. 36 (1997) 2207–2218.

[133] A.E.H. Daifullah, S. El-Reefy, H. Gad, Adsorption of p-nitrophenol on inshas incinerator ash and on the pyrolysis residue of animal bones, Adsorp. Sci. Technol. 15 (1997) 485–496.

[134] Z. Aksu, A. Calik, A comparative study of the biosorption of iron (III)-cynade complex anions to Rhizopus arrhizus and Chlorella vulgaris, Sep. Sci. Technol. 34 (1999) 817–832.

[135] M.M. Saleh, A.A. Atia, Removal of some surfactants from dilute aqueous solutions using charcoal, Adsorp. Sci. Technol. 17 (1999) 53–64.

[136] M.S. El-Geundi, I.H. Aly, Equilibrium studies during the adsorption of acid dyestuffs onto maize cob, Adsorp. Sci. Technol. 9 (1992) 121–129.

[137] S.A. Darwish, K.A. Halhouli, N.M. Al-Dhoon, Adsorption of phenol from aqueous systems onto spent oil shale, Sep. Sci. Technol. 31 (1996) 705–714.

[138] S.C. Kwon, D.I. Song, Y.W. Jeon, Adsorption of phenol and nitrophenol isomers onto montmorillonite modified with hexadecyltrimethylammonium cation, Sep. Sci. Technol. 33 (1998) 1981–1998.

[139] I. Safarik, M. Safarikova, Copper phthalocyanine dye immobilized on magnetite particles: an efficient adsorption for rapid removal of polycyclic aromatic compounds from water solutions and suspensions, Sep. Sci. Technol. 32 (1997) 2385–2392.

[140] Z. Feleke, Y. Sakakibara, Nitrate and pesticide removal by a combined bioelectrochemical/adsorption process, Water Sci. Technol. 43 (2001) 25–33.

[141] G. Sekaran, K.A. Shanmugasundaram, M. Mariappan, K.V. Raghavan, Utilisation of a solid waste generated in leather industry for removal of dye in aqueous solution. Ind. J. Chem. Technol. 2 (1995) 311–316.

4 Water Treatment by Reverse Osmosis Method

CHAPTER OUTLINE

4.1 Introduction

Reverse osmosis (RO), also known as hyperfiltration, is a classical method of purification and came into existence since the advancement of effective semipermeable membranes [1], but it has achieved a great attention now a days as the technique of choice for the removal of anions, cations, and other soluble organic and inorganic materials from wastewater, saline water, leachates, etc. The separation and concentration of the dissolved species are achieved due to the hydraulic gradient across the semipermeable membrane. A pressure greater than the osmotic pressure (due to dissolved species) is applied for the process in an operation. RO has been used as a separation and concentration technique at macro and microlevels for the removal of large, nonpolar, and ionic compounds [2]. It is found to be more cost effective, for the removal of toxic substances, than other conventional techniques, like activated carbon

Environmental Water, http://dx.doi.org/10.1016/B978-0-444-59399-3.00004-0

adsorption. The economy of the method for a variety of industrial wastewaters has been discussed by Fosberg et al. [3]. The effectiveness and the application of RO on large scale are based on the fact that total dissolved solids (TDSs), organic dissolved matter, etc., were removed up to 85–99% [4,5]. Due to the wide range of applications of RO, it has been used for treatment of wastewater in petrochemicals [6–8], electroplating [9], textiles [10], coal and gasification [11], pulp and paper [12], steel [5], electronic wastewaters [13], etc. The removal of 15 major pesticides from aqueous industrial waste, containing chlorinated hydrocarbon and organophosphorous compounds [14], and toxic metals and cyanides [15] has been demonstrated. Several membrane materials have been used successfully to treat electroplating rinse water, with high oxidation potential and at extreme pH levels [16]. Varying degrees of rejection of specific organic species have been studied by Gaddis et al. [10] in textile effluent with high level of toxic substances. Effectiveness of RO for the treatment of sanitary waters [17] and municipal leachates [18,19] has been evaluated. RO has also been used very frequently [20–24] to remove dissolved organic matters (DOMs) from surface waters. Reports on the removal of DOMs (with very high dissolved organic carbon [DOC] and very low inorganic solutes) from black water streams are also available [24]. Besides, the rejection of bacteria, virus, and other microbes are 100% and, therefore, is used in the preparation of ultra pure water for pharmacy, medicines, and electronics. RO is today's most economic process for saline water treatment to provide potable water [25].

4.2 RO Membrane

The most important part of RO process is semipermeable membrane, which is made of different materials. Various types of semipermeable membranes have been used for RO process [25–29], but the most common ones are of cellulose, nylon, polyether, polyethyl urea, polyphenyl oxides, phenylenes [5,25,30], and polyamide [31]. Membranes are normally casted from a ternary solution of cellulose acetate, acetone, and formamide according to well-known method [32,33]. To achieve the required filtration, membranes are arranged in tubular, disc, plates, spiral, and hollow fiber forms. It is clear from literature that tubular RO can successfully be applied for treatment of coal gasification liquors for chemical oxygen demand (COD), TDS, and nitrogen ammonia removal. Tubular devices have been found for commercial use in special applications, especially for treatment of effluents, recovery of chemicals from process streams, etc. [34]. However, tubular and disc tubes are successfully coupled for the treatment of municipal leachates for TDS, COD, and ammonia–nitrogen removal. The size of tubular or disc RO plant, pressure, pH of feed, etc., are controlled according to the water treatment requirements [5]. The working efficiency of RO membrane is expressed in terms of percentage rejection of solutes and expressed as follows:

$$\mathrm{PR} = \frac{C1 \times C2 \times 100}{C1} \tag{4.1}$$

or

$$E = \frac{C2}{C1}, \tag{4.2}$$

where PR is the percentage rejection, *C*1 is the molar concentration of solute in feed water, *C*2 is the molar concentration of solute in permeate (filtrate), and *E* is the membrane working efficiency.

4.2.1 Mechanism of Filtration

Anderson [35] described the mechanism of RO on the basis of partition coefficients of solutes between water and membrane, while Krasen and Eiseman [36] and Kamizarva and McCutchan [37] explained the selectivity of RO membranes on the basis of relative free energy of interaction between water and membrane sites. The other workers [4,38–41] have also described the mechanisms of RO that is more acceptable. According to Duvel and Helfgott [38], the percentage of rejection is directly proportional to the number of carbon atoms or the molecular weights of organic/inorganic species, but this was not found to be the complete description because of the geometry of the molecule, i.e. steric configuration and branching overshadow the former properties. In order to determine the relative effect of molecular weight and molecular size, a series of alcohols was studied in which the degree, position of branching, and the position of the hydroxyl groups were varied. It has been observed that the percentage of the rejection increased as the molecular weight or the size of the molecule increased. The hydrogen bonding, the process of sorption and/or dissolution, and diffusion has been considered responsible for the membrane permeation. Phenols, amines, amides, etc., form H-bonding to the cellulose membrane and get passed through it. It is due to H-bonding that such molecules pass through the membrane, while the other ions and organic species get rejected. Based on the experiments, it has been postulated that in cellulose acetate membranes, random thermal motion of the polymer segments plays an important role in opening the gaps or space between adjacent segments and help in the membrane filtration process.

4.3 Membrane Flux

The fluxes of the solutes and solvents in RO have been described by various systems of equations [42]. The membrane flux and the mass transfer have also been explained by a series of other equations by Slater et al. [4]. In desalination and wastewater treatment processes, the solvent or permeate flux has been referred to the water and defined as the volume of permeate flux produced per unit membrane area for a time interval and is given by the following equation:

$$\mathrm{MF} = \frac{\mathrm{VP}}{\mathrm{SA}t}, \tag{4.3}$$

where MF is the membrane flux, VP is the volume of permeate over time *t*, and SA is the surface area of the membrane.

At a given pressure, the rate of membrane flux decreases with the increase of ionic strength (Sun et al., 1995) and it also decreases over a long period of time. The rate of flux also varies by varying the pressure (increase by $1.95 \times 10^{-2}\ m^3/m^2d$ per $100\ kN/m^2$ pressure increase) and temperature (increase by $0.012\ m^3/m^2d$ per 10°C temperature increase). The membrane flux also increases by increasing the cross-sectional area of the membrane.

4.4 Fouling and Clogging of Membranes

The life of RO membranes has been considered to be 2-5 years depending upon the nature of the wastewater to be treated. The flux and the quality of the permeate may decrease over a long period of time due to membrane fouling [43,44]. Fouling has been defined by Eykamp [45] as a phenomenon in which the membrane undergoes plugging or coating by certain elements in the feed stream. The flux is reduced as the membrane surface commences. Dissolved silica [46] is the major problem for RO membranes as it closes the membrane pores quickly. Amorphous silica, sulphates, and siliceous precipitates get deposited inside the membranes and clog them easily. After a long period of time, humic acids, bacterial slimes, or hardness scales may accumulate on the RO membranes. Phenols have been found to clog the membranes. Fouling of membranes has been reported by the sanitary landfill and copper plating solutions [47].

4.5 Precautions and Cleaning of the Membranes

The cleaning solutions and other instructions recommended by different manufacturers for the long life of RO membranes have been described [25,48]. The pretreatment is necessary to minimize the concentration of colloidal dispersion and to increase the efficiency and life of RO systems [49]. Physicochemical coagulation has been used [50] to minimize colloids and turbidity. Coagulation with lime has been recognized for its utility in treatment of industrial wastewater. Coagulation was [4] found effective in removing turbidity containing dispersed oil phases, metal ions, and suspended matters in landfill leachates. Pretreatment to RO systems has been applied successfully on commercial scale to prevent fouling in several treatment plants [51]. Sodium hydroxide solution (pH 9–11) has been used to clean the RO membranes in case of silica and sulphate fouling [25,46]. In commercial applications, silica is removed by passing the raw water through an acrylic or styrenic (Amberlite IRA series) strong base resin [9,52]. Besides, silica can also be removed from membranes by ion retarding resins [24] that has a high affinity for strong acids together with conversion of the weak acid [Si(OH)4] into much stronger acid (H2SiF6). The bacterial inhibitor solutions are circulated (to check the bacterial growth) in to the RO tubules or discs before stopping the process for a long period [46]. In addition to this, Biotex (0.5%), citric acid (2%, pH 4.5), ethylenediaminetetraacetic acid EDTA) (0.5%), and

P3 Ultrasil 50 (0.5%) were circulated at reduce pressure for 1 h followed by a tape water rinse for 30 min for cleaning the RO systems [5]. Phenolics may be removed from RO membranes by circulating hydrogen peroxide solution.

4.6 Factors Controlling Filtration

The filtration by RO is controlled by a number of factors. The most important are pH, pressure, size and the molecular weight of the solute, time of operation, etc. The pH is one of the most important parameter affecting the RO filtration process. Sun et al. [46] conducted laboratory experiments to examine the effect of pH of the percentage rejection of a low molecular weight organic acid (0.006 M) in the range 4–10. It was observed that the percentage rejection of acetic acid (0.006 M) increased up to pH 8.0 and then became constant. The effect of pH on the percent rejection of HCl, NaCl, and sodium acetate was also studied by the same authors and it was found that sodium chloride and sodium acetate were strongly rejected while the HCl was least rejected. The chromate rejection increased from 5.5 to 8.0 pH and then became constant [53]. Moreover, semipermeable membranes work efficiently under a specific range of pH as cellulose acetate, cellulose triacetate, and polyamide membranes may work properly at 2.5–7.0, 2–9, and 2–10.5 pH ranges, respectively. The effect of pH on the solute rejection may be explained on the basis of H-bonding between the solute and the semipermeable membrane.

The effect of the operating pressure on the flux in RO has not been studied in greater detail. It was observed by Chian and Fang [40] that the rejection of some low molecular weight polar compounds increased with the increase of pressure. However, no effect on the permeate quality has been observed by the increase of pressure [22]. The preferential separation of water or solute at membrane–solution interface might explain these different observations. However, the RO system has been run in both steady and unsteady states with varying pressure. A pressure of 690–2760 kN/m^2 was used [4] with 6.3×10^{-5} m^3/s as the membrane feed rate. Sun et al. [46] and Schoeman et al. [5] have used 150–200 psi and 4000 kPa pressures, respectively. From the results obtained by various workers, it was found that the rate of flow varied with the feed pressure and is directly proportional to the pressure applied and the permeate flux increases by a value of 1.95×10^{-2} m^3/m^2d per 100 kN/m^2 pressure. The percentage rejection of a solute is also governed by the size and molecular weight of the solute. Duval and Helfgott [54] have carried out an extensive study on the size and molecular weight of a series of aliphatic alcohols with respect to its percentage rejection. They have reported that the percentage rejection increases as the molecular size and molecular weight increases. The rejection of solute of high molecular size and weight is explained on the basis of sorption and/or dissolution. Various workers have studied the variation in the rate of filtration, permeate flux, rejection, etc., during the course of time and observed no marked effect on the filtration in spite of the long use of RO membranes. However, the rate of filtration slightly decreases after a long period of use. Schoeman et al. [5] has observed that permeate flux decreases up to 173 L/m^2d from 300 L/m^2d after an operation of 950 h. The conductivity

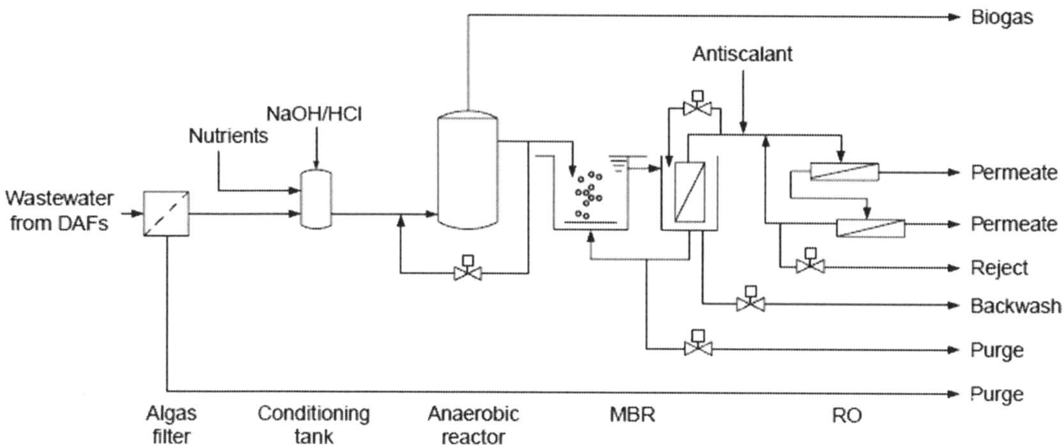

FIGURE 4-1 An anaerobic reactor followed by a MBR and a final RO.

and COD were also found almost constant even after a long period of use. A typical anaerobic reactor followed by a membrane bioreactor (MBR) and a final RO system is shown in Figure 4-1.

4.7 Applications of RO

RO has been used for the removal of various pollutants and the application of this technology is described in the following subsections.

4.7.1 Organic Pollutants

Drewes et al. [55] reported that microfiltration (MF) followed by RO and soil-aquifer treatment (SAT) are the two principal technologies considered for indirect potable reuse of wastewater. The study was conducted at the Northwest Water Reclamation Plant, Mesa (Arizona), evaluated MF/RO and SAT (>6 months residence time) treated tertiary effluent with respect to organics removal. The study revealed that final total organic carbon (TOC) concentrations of MF/RO and SAT were 0.3 and 1.0 mg/L, which resembled the character of natural organic matter present in drinking water. Kang et al. [56] described reused of water in textile dying processes after treatment of textile effluents by H_2O_2/ultraviolet (UV) oxidation combined with RO separation. The conductivity of textile effluents was from 2340 to 4560 μs/cm. Addition of auxiliary chemicals used during the dyeing processes increased the conductivity in textile wastewaters. The H_2O_2/UV pre-oxidation of textile effluents mineralized or oxidized DOC effectively. However, the removal of conductivity and hardness was poor. Pretreatment of the textile effluent by H_2O_2/UV oxidation decreased silt density index (SDI) values and osmotic pressure and increased permeate flux followed by RO separation. H_2O_2/UV pre-oxidation and RO posttreatment improved the textile effluent quality and was able to meet the water quality criteria for water reuse in the textile industry. Chen and Chen [57] presented an RO method of water treatment in thin film transistor liquid crystal display (TFT-LCD) industry. TFT-LCD industrial wastewater was recycled by integrated process of MBR and RO. Ultrafiltration (UF) membrane device was incorporated with biological reactor to attain high-quality effluent and for the good removal performance. A good water quality of permeate was obtained with 5 mg/L, 2.5 mg/L, and 150 μs/cm for COD, TOC, and conductivity, respectively. The treated water was recycled for the cooling tower and other purposes. Lee and Lueptow [58] reported clean water supply by RO using stirred cell filtration experiments. Low pressure reverse osmosis membranes were used to obtain high flux of permeate as well as high rejection. The removal of detergent was 99% and that of dissolved salt above 90% in single-pass treatment. On the other hand, TOC removal was nearly 80%. The removal efficiency for nitrogen compounds was also improved by increasing the shear rate near membrane surface. The product water in two passes could meet the hygiene water requirements for human space missions, and the product water in three passes could meet potable water regulations with overall recovery of 77%. Downing et al. [59] reported that the increase in salt concentration that occurs with each water use is the major limitation with wastewater reclamation and reuse scheme. Integrated use of wastewater pond design, solids separation equipment, and membrane technology removes salt in a more economical way. Effluent from an Advanced System Waste Water Pond System (AIWPS) Facility was clarified in a Krofta Supracell dissolved air flotation unit and a slow sand filter prior to final treatment in an expertise RO unit. The ponds of the AIWPS Facility removed an average of 82% of soluble biochemical oxygen demand (BOD) and 80% of soluble nitrogen. Following clarification, filtration, and RO treatment, the pollutant removals were >99% for soluble BOD, >99% for soluble nitrogen, and 98% for TDS. Based on membrane fouling rate data, the cleaning interval for the RO membranes in a full-scale AIWPS-RO Facility would be over 100 days. This interval is at par with that typically seen in full-scale reclamation facilities treating secondary activated sludge effluent with MF prior to RO. Srinivasan et al. [60] reported the removal of BOD and COD via a combination of anaerobic treatment and RO for recyclable water from the effluent. The reactor was operated under different organic loading rates and hydraulic residence times at a constant feed pH of 7.0. The COD, BOD, total alkalinity, pH, and total volatile acids (TVA) were monitored in the samples drawn from different sampling ports provided in the reactor. Separation performance was evaluated at varying feed pressure (11–41 bar) and found to improve with increasing pressure. High rejection of dissolved solids (approximately 98%), COD, BOD, and almost complete removal of color were achieved with reasonable flux and water recovery.

4.7.2 Oils

Murray-Gulde et al. [61] reported that the hybrid RO-constructed wetland treatment systems presented a viable alternative for treatment and reuse of waters from oil fields.

The produced water was used for irrigation purposes. The pilot scale RO-constructed wetland system effectively decreased conductivity by 95% and TDS by 94% in the brackish produced water tested in this study. Specifically, the performance of the hybrid treatment system was examined in terms of outflow water conductivity, TDSs, and toxicity using Ceriodaphnia dubia and Pimephales promelas in a 7-day static/renewal exposure tests. Yu et al. [62] used an integrated MF-RO for the removal of vegetable oil industrial effluent. According to authors, in MF continuous process, the steady flux was around 90 $L/m^2/h$ when the concentrated multiple reached 16, and the oil content in permeate was less than 12 mg/L. In RO continuous process, antifouling membrane was used to treat permeate from the ceramic membrane process in order to improve water quality. RO process had a permeate flux of 24 $L/m^2/h$ and water recovery rate of 95%. The permeate from the RO stage was free of oil, and its TOC and conductivity were less than 0.6 mg/L and 50 μs/cm, respectively. The results demonstrated that the two-stage membrane process combining MF and RO was highly efficient in the treatment of oil-containing wastewater.

Steinle-Darling et al. [63] proposed RO as effective treatment method for a wide range of organics. *N*-nitrosodiethylamine (NDEA), *N*-nitrosomethylethylamine (NMEA), *N*-nitrosodipropylamine (NDPA), *N*-nitrosodibutylamine (NDBA), *N*-nitrosopyrrolidine (NPyr), *N*-nitrosopiperidine (NPip) were the organics studied. Generally, small, polar, uncharged molecules such as *N*-nitrosodimethylamine (NDMA) were poorly rejected. Tepus et al. [64] described nanofiltration (NF) and RO membranes along with the use of two adsorption resins; one derived from styrenedivinylbenzene and another from polystyrene; as effective methods in removing pollutants such as nitrate and pesticides (atrazine, deethylatrazine, and deisopropylatrazine) from actual groundwater in Slovenia. The RO membrane displayed a high rejection of all compounds, while NF showed lower compound rejection hence suitable for atrazine removal. Ben Amar et al. [65] reported that the wastewater of dyeing factory can be recycled and reused by coupling the activated sludge treatment with either NF or RO. For the study of the NF/RO textile wastewater treatment, experiments were conducted with spiral wound membranes in order to be closest to the industrial configuration. After analyzing the removal efficiencies of suspended solids and COD of the treatment plant, NF was conducted using an HL2514TF spiral wound membrane preceded by UF treatment. According to authors, the results showed that NF allowed higher yield while respecting the Tunisian standard of water reuse (COD < 90 mg/L). Above 9 bar, TDS rejection reached 60% and the hardness was lower than the factory constraint (100 mg/L of $CaCO_3$), allowing the reuse of the water in the process. You and Teng [66] described the removal of azo and Reactive Black 5 dyes from textile industrial effluent by using combined aerobic membrane bioreactor(AOMBR)/RO processes. The results showed that in the anaerobic sequencing batch reactor RB5 dye was degraded to form aromatic amine intermediate metabolites, which were further mineralized in the AOMBR. Five anaerobic high-dye-degrading bacteria were isolated, which were identified to be the same species of *Lactococcus lactis* by 16S ribosomal RNA sequencing. Kegel et al. [67] used RO and activated carbon filtration for the removal of organic micropollutants from river bank filtrate. Total removal efficiency of 47 organic micropollutants was investigated. Very high removal efficiencies for almost all organic micropollutants by the subsequent activated carbon, fed with the permeate stream of the RO element, were observed except for the very small and polar NDMA and 1,4-dioxane. RO and subsequent activated carbon filtration were complementary and their combined application results in the removal of a large part of these emerging organic micropollutants. Park and Hu [68] described the removal of inorganic, organic, and bacterial contents. According to authors, RO membranes improved water quality of drinking water in terms of organic matter (microbes). More than 99% of bacterial cells in the tap water was retained by the RO membranes, leaving <50 cells/mL in the permeate water.

4.7.3 Pharmaceuticals

Gabet-Giraud et al. [69] removed estrogenic hormones and beta blockers in rural and urban wastewater treatment plant (WWTP) located in France by using ozonation, RO, and activated carbon filtration. Biological treatments are efficient to remove estrogens from the dissolved phase, with removal rate of around 90%. For beta blockers, acebutolol and nadolol are efficiently removed (mean removal rate of 80%), whereas sotalol and propranolol are hardly impacted by biological treatments. Theepharaksapan et al. [70] removed Dibutyl phthalate (*DBP*), di-(2-ethylhexyl) phthalate (*DEHP*), and biphenyl up to 100%, 99.6%, and 98.0% by chemical coagulation, sand filtration, MF and RO. Boleda et al. [71] reported that removal of drugs and pharmaceuticals at drinking water treatment plant (DWTP). DWTP has a common treatment consisting of dioxychlorination, coagulation/flocculation, and sand filtration and then water is split in two parallel treatment lines: conventional (ozonation and carbon filtration) and advanced (UF and RO) to be further blended, chlorinated, and distributed. Full removals were reached for most of the compounds. Iopromide (up to 17.2 ng/L), nicotine (13.7 ng/L), benzoylecgonine (1.9 ng/L), cotinine (3.6 ng/L), acetaminophen (15.6 ng/L), erythromycin (2.0 ng/L), and caffeine (6.0 ng/L) with elimination efficiencies $\geq$94% were the sole compounds found in the treated water. The advanced treatment process showed a slightly better efficiency than the conventional treatment to eliminate pharmaceuticals and drugs of abuse. Sui et al. [72] removed 13 pharmaceuticals, antibiotic, antilipidemic, anti-inflammatory, antihypertensive, anticonvulsant, stimulant, insect repellent, and antipsychotic in four WWTPs of Beijing, China. These pollutants were eliminated by sand filtration, but the ozonation and MF/RO processes employed in two WWTPs were very effective to remove them, showing their main contributions to the removal of such micropollutants in wastewater treatment. Most of the target compounds were detected, with the concentrations of 4.4–6.6 μg/L and 2.2–320 ng/L in the influents and secondary effluents, respectively. Kimura et al. [73] investigated the influence of organic macromolecules on removal of six pharmaceuticals by NF/RO membranes (UTC-60 and LF10). A high-pressure-driven membrane such as an NF membrane and a RO membrane were effective in controlling

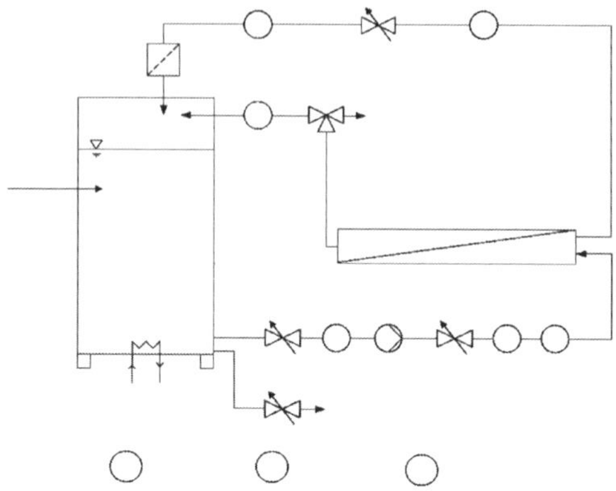

FIGURE 4-2 An experimental design of the pilot plant for the removal of pharmaceutical residues by NF and RO.

pharmaceuticals in wastewater treatment. Six pharmaceuticals were removed by NF/RO membranes (UTC-60 and LF10). As per the authors, the characteristics of the organic macromolecules contained in the wastewater effluents differed depending on the type of treatment, implying that removal of pharmaceuticals by NF/RO membranes is influenced by the type of pretreatment employed. An experimental design of the pilot plant for the removal of pharmaceutical residues by NF and RO is given in Figure 4-2.

4.7.4 Inorganic Pollutants

Di Palma et al. [74] reported that the contamination of lead and copper from the soil can be treated by soil flushing, membrane treatment, acidification, and metal precipitation. The membrane treatment was done to concentrate the solutions extracted by using EDTA, 75% of the water was reduced, and a permeate metal concentration which in compliance with the Italian Environmental Regulation was obtained. Topp et al. [75] described RO system for treating water from uranium mine mill. The authors reported good water quality suitable for release to the environment. The unique features of the proposed process were recycling of the lime/CO_2 softening sludge to the uranium mill as a neutralant, use of power station off-gas for carbonation, application of residual ammonia as pH buffer in carbonation, and recovery and recycling of ammonia from the RO reject stream. Walker et al. [76] reported the application of RO system for removal of arsenic from groundwater of Lahontan Valley in rural Nevada, United States. The results indicated that RO systems removed arsenic up to 80.2% from well water. Chang et al. [77] described the use of secondary effluents reclaimed water in agriculture depends on its salinity and sodium adsorption ratio (SAR). Membrane technologies, especially NF and RO, have played a key role reclaiming the secondary effluent. RO removed monovalent and divalent cations simultaneously. NF process rejected preferably divalent cations and most monovalent ions. The primary objective of this study was to examine SAR values of the reclaimed water after NF and RO processes. The measured SAR values of the secondary effluent, NF permeate, and RO permeate were 1.78, 4.67, and 0.72, respectively. The SAR value after NF (4.67) increased more than twice that of the feed solution, whereas the SAR of the RO permeates decreased to 0.72. Voorthuizen et al. [78] reported NF and RO to recover the nutrients from anaerobically treated black water. A large number of commercial tubular, capillary, and flat sheet NF and RO membranes was tested on laboratory scale on their performance to meet the Dutch guidelines for ammonium and phosphate. The ammonium and phosphate concentrations used were based on the effluent composition of anaerobically treated black water. Ammonium and phosphate rejection were both measured in synthetic single salt and multi-ion mixtures and in anaerobic effluent. The rejection for ammonium (30–95%) was neither sufficient for discharge nor potable water use. The rejection of phosphate (74–99%) is in most cases almost sufficient to meet the standards for potable water.

4.7.5 Biological Pollutants

Ganzi and Parise [79] reported that RO was found to be effective for the removal of trace ionic contaminants. Furthermore, they reported that RO systems can be used for water with low levels of bacteria and pyrogens. López-Ramírez et al. [80] reported physical–chemical pretreatment for the RO unit for reclaiming secondary effluents. Fouling and biofouling of the cellulose–acetate RO membranes were prevented using tertiary treatments units. The optimization of pretreatment involved application of various concentrations of lime to raise the pH to 10.3–12.1 and to stabilize the sludge generated as well as different dosages of ferric chloride (15, 20, and 25 mg/L) for the coagulation and solid–liquid separation. Sodium hypochlorite (8 mg/L) and UV disinfection are used for microbiological control. The water quality obtained, under the optimum conditions (pH = 10.5, $FeCl_3$ = 25 mg/L, anionic flocculant = 0.5 mg/L, sodium hypochlorite = 8 mg/L), was high, showing an average conductivity of 66 μS/cm and low COD values of 4 mg O_2/L. The product water was suitable for injection into a groundwater aquifer to counteract seawater intrusion. Xu et al. [81] reported that for drinking water treatment and wastewater reclamation, RO, NF, and ultra low pressure RO membranes were best methods for removing organics micropollutants. Comerton et al. [82] assessed the potential reuse applications of municipal wastewater via a MBR-RO. The authors reported the removal of waterborne pathogens and nitrate by RO. By using MBR-RO system, Jones et al. [83] reported the removal of endocrine disrupting compounds and related micropollutants from wastewater. A combination of granular activated carbon, ozone treatment, membrane filtration, and RO are the advance treatment techniques proposed in the United Kingdom. But high cost, increased energy consumption and CO_2 emission are its associated drawback.

4.7.6 Direct Applications

Qin et al. [84] described dual membrane UF/RO processes for treating a combined liquor of spent alkaline, acidic, and nickel-plating rinses, which resulted in a treated water of a quality suitable for reuse as substitute for town water for the purpose of rinsing. UF pretreatment combined with an RO membrane unit has been selected as the treatment system by the authors. The pilot plant had successfully operated for 6 months, consistently produced high-quality water (<95 μS/cm) at an overall water recovery of 67.5%. The quality of reclaimed water was better than town water used at the factory. Thörneby [85] reported that RO is one of the promising methods for leachate treatment as it reduces the pollutant discharge. RO technology was tested at a pilot plant at Hedeskoga landfill in southern Sweden. This landfill receives municipal solid waste and industrial, commercial, and institutional waste and produces about 3000 m^3/ha of leachate annually. Hasar et al. [86] reported that RO was an advanced treatment method for the leachate with high organic and ammonium concentration. The RO was applied to the collected aerobic/anoxic MBR. RO provided high-quality effluent by reducing the effluent COD from MBR to less than 4.0 mg/L at solid retention time (SRT) of 30 days. Bartels et al. [87] reported improved process design and membrane for the treatment of municipal and industrial wastewaters. Factors that played key roles in the use of RO membranes included ultra or MF pretreatment, low fouling membranes, flux rate, and recovery and control of fouling and scaling. In particular, high flux rates were used when UF or MF pretreatment is used. Chu et al. [88] described the treatment of wastewaters of several WWTPs of industrial parks in Taiwan. The treatment process included a sand filter, an UF unit, and a RO module. The results of this study demonstrated that the production quantity and quality were stable and appropriate for various purposes, including both industrial and domestic applications. Diltz et al. [89] described single cartridge prototype RO system for post-treatment of an anaerobic fermentation broth (ethanol, butanol, acetic acid, oxalic acid, lactic acid, and butyric acid). As per authors, the results were encouraging and useful for further studies on the feasibility of incorporating RO technology into an integrated and field deployable wastewater management and water recovery system. Correlations of the

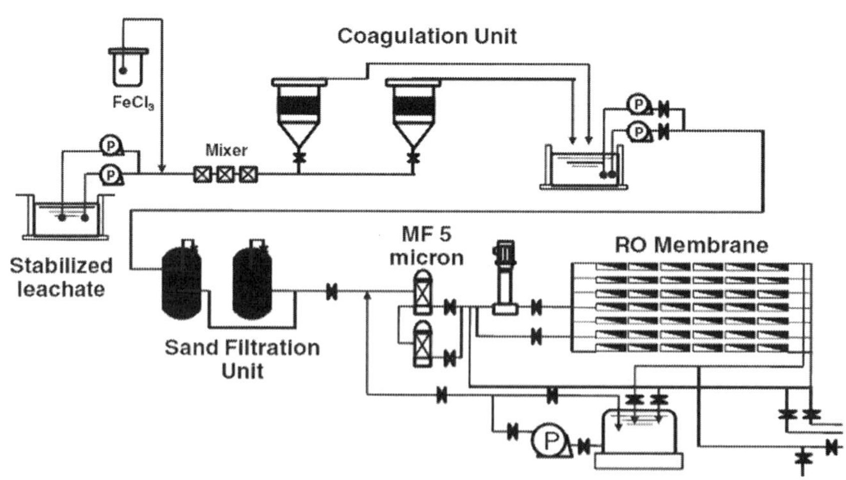

FIGURE 4-3 Schematic diagram of advanced leachate treatment system.

rejection and recovery efficiencies for these organic species, individually and in simulated mixtures, were obtained as a function of feed pressure with and without recirculation of the retentate. The actual fermentation broth obtained from a continuous-flow bio-hydrogen process was treated by the RO system under the operating conditions similar to those used in the baseline tests, resulting in greater than 95% removal of TOC. Ordóñez et al. [90] reported a pilot scale anaerobic and aerobic treatment followed by UF and RO for treating paper mill wastewater. According to authors, the quality of treated water was quite good that might be replaced fresh water in some applications. A schematic diagram of advanced leachate treatment system is shown in Figure 4-3.

Conclusions

RO has been used as a technique of choice for the separation, concentration, and desalination of organic and inorganic species as the ionic species rejection was found to be highly effective (up to 99%). The system has also been recognized as most effective and efficient. Based on favorable attractive operating characteristics, the technique enjoys wide acceptance in the preparation of potable and ultra pure waters from saline, surface, and groundwaters but have certain limitations for the purification of wastewaters. The pretreatment of (filtration by microfilters, coagulations, etc.) wastewater is a costly, time-consuming, and difficult affair, which make RO as a costly process. The feed water for RO should be free from turbidity, oil, grease, etc. Besides, semipermeable membranes work under some specific range of pH. The fouling of membranes reduces the life of the membranes and consequently results in decline in RO permeate flux. Because of these limitations, RO could not find vast application in wastewater purification.

References

[1] C.E. Reid, E.J. Breton, Water and ion flow across the cellulose membranes, J. Appl. Polym. Sci. 1 (1959) 133–134.

[2] W.G. Light, in: W.J. Copper (Ed.), Removal of chemical carcinogens from water/wastewater by reverse osmosis in Chemistry in Water Reuse, vol. 1, Ann Arbor Science, Ann Arbor (MI), 1981, p. 352.

[3] T.M. Fosberg, D. Mukhopadhyay, T.M. O'Neal, R.C. Whalen, Selecting energy efficient wastewater treatment systems, Chem. Eng. Prog. 77 (1981) 63–67.

[4] C.S. Slater, R.C. Ahlert, C.G. Uchrin, Treatment of landfill leachates by reverse osmosis, Environ. Progr. 4 (1983) 251–256.

[5] J.J. Schoeman, A. Steyn, P.J. Scurr, Treatment using reverse osmosis of an effluent from stainless steel manufacture, Water Res. 30 (1996) 1979–1984.

[6] M.A.C.A. Zeitoun, C.A. Roorda, G.R. Power, Total Recycle for Petrochemical Waste Brine Containing Refractory Contaminants. EPA-600/2-79-021, US Environmental Protection Agency, Cincinnati (OH), 1979, pp. 101–125.

[7] N.G. Muratova, V.B. Shevchenko, G.F. Stychinskii, A.V. Dolganova, I.N. Smirnov, Study of waste-water treatment by reverse osmosis, Int. Chem. Eng. 19 (1979) 350–352.

[8] R. Matz, E. Zinster, Herscovici G., Pressure driven membrane process in the treatment of industrial effluents, Desalination 24 (1978) 113–128.

[9] J.T. McNulty, Anion exchange resin kinetics in mixed bed condensate polishing, in: D. Naden, M. Streat (Eds.), Ion Exchange Technology, Ellis Norwood, London, 1984, pp. 50–65.

[10] J.L. Gaddis, H.G. Spencer, Separation of Toxic Substances by Hyper Filtration Symposium Proceedings: Textile Industry Technology. EPA-600/2-79-104, US Environmental Protection Agency, Cincinnati (OH), 1979, pp. 107–114.

[11] A.R. Williams, The use of reverse osmosis for the purification of coal gasification liquors, in: M.K. Turner (Ed.), Effective Industrial Membrane Processes: Benefits and Opportunities, Elsevier science Publishers Ltd, Barking, Essex, 1991.

[12] A.J. Wiley, L.E. Dambruch, P.E. Parker, S.H. Dugal, Treatment of bleach plant effluents, TAPPI 61 (1978) 77–80.

[13] M.D. Beasley, L. Luttinger, C. Caprio, Reverse Osmosis Provides Reusable Water from Electronic Waste. Proceedings Of the 32nd Industrial Waste Conference, 1977, Purdue University, West Lafayette (IN), 1977, pp. 630–638.

[14] E.S.K. Chian, W.N. Bruce, H.H.P. Fang, Removal of pesticides by reverse osmosis, Environ. Sci. Tech. 9 (1975) 52058.

[15] L.J. Kosarek, Removal of various toxic heavy metals and cyanides from water by membrane processes, in: W.J. Cooper (Ed.), Chemistry in water reuse, vol. 1, Ann Arbor Scienc, Ann Arbor (MI), 1982, pp. 261–280.

[16] K.J. McNulty, P.R. Hoover, Evaluation of Reverse Osmosis Membranes for Treatment of Electroplating Rinse waters. EPA-600/2-80-084, US Environmental Protection Agency, Cincinnati (OH), 1980, p. 5.

[17] E.S.K. Chian, F.B. De Walle, Evaluation of leachate treatment. EPA-600/2-77-186b, in: Biological and Physical-Chemical Processes, vol. II, US Environmental Protection Agency, Cincinnati (OH), 1977, p. 266.

[18] T.A. Peter, Treatment of landfill leachates by conventional and high pressure reverse osmosis based on the disc tube module, in: Engineering of Membrane Processes 11, Environmental Applications, vol. 11, Ciocco, Tuscany, Italy, 1994.

[19] B. Weber, F. Holz, Landfill leachates treatment by reverse osmosis, in: M.K. Turner (Ed.), Effective Industrial Membrane Processes: Benefits and Opportunities, Elsevier Science Publishers Ltd., Barking, Essex, 1991.

[20] T.A. Clair, J.R. Kramer, M. Sydor, D. Eaton, Concentration of aquatic dissolved organic matter by reverse osmosis, Water Res. 25 (1991) 1033–1037.

[21] J.S. Kim, E.S.K. Chian, F.M. Saunders, E.M. Perdue, M.F. Giabbai, Characteristic of humic substances and their removal in water treatment, in: I.H. Suffet, P. MacCarthy (Eds.), Aquatic Humic Substances: Influence on Fate and Treatment of Pollutants (Advances in Chemistry Series), vol. 219, American Chemical Society, Washington, DC, 1989, pp. 473–497.

[22] H. Odeggard, S. Koottatep, Removal of humic substances from natural waters by reverse osmosis, Water Res. 16 (1982) 613–620.

[23] H. Odeggard, T. Thorsen, Removal of humic substances by membrane processes, in: I.H. Suffet, P. MacCarthy (Eds.), Aquatic Humic Substances: Influence on Fate and Treatment of Pollutants, Adv. Chem. Ser., vol. 219, Am. Chem. Soc., Washington, DC, 1989, pp. 769–782.

[24] S.M. Serkiz, E.M. Perdue, Isolation of dissolved organic matter from the Suwannee River using reverse osmosis, Water Res. 24 (1990) 911–916.

[25] W.H. Pohland, Reverse osmosis, in: W. Lorch, 1981 (Eds.), Hand Book of Water Purification, second ed., Ellis Hoorwod Ltd., England, 1981, pp. 316–372.

[26] M. Deinzer, R. Melton, D. Mitchell, Trace organic contaminants in drinking water, their concentration by reverse osmosis, Water Res. 9 (1975) 799–805.

[27] M. Reinhard, N.L. Goodman, D.C. McCarthy, D.G. Argo, Removing trace organic by reverse osmosis using cellulose acetate and ployamide membranes, J. Am. Wat. Works Assoc. 78 (1986) 163–174.

[28] J.H. Baier, B.W. Lykins Jr., C.A. Frank, C.J. Kramer, Using reverse osmosis to remove agricultural chemicals form ground water, J. Am. Wat. Works Assoc. 79 (1987) 55–60.

[29] S.C. Lynch, J.K. Smith, Evaluation of reverse osmosis to concentrate organic contaminants from water, in: Advance in chemistry series, vol. 241, American Chemical Society, Washington, DC, 1987, pp. 431–465.

[30] S. Loeb, S. Sourirajan, New Water Desalting Process Developed at UCLA, University of California, Office of Public Information Press Release, 1959.

[31] C.F. Schutte, G. Belfort, Rejection of alkyl phenols by membranes, Water Sci. Tech. 19 (1978) 967–979.

[32] S. Mannjikian, Semipermeable desalination membranes from organic casting solutions, University of California, Los Angeles, 1965. Department of Engineering Report No. 65-13.

[33] H. Yasuda, Composite reverse osmosis membranes prepared by plasma polymerization, in: S. Sourirajan (Ed.), Reverse Osmosis and Synthetic Membranes, NRC Canada, Ottawa, Ontario, 1977.

[34] D. Pepper, Effluent Treatment by Reverse Osmosis, Conference on Pollution Control, Amsterdam, Netherlands, 1974.

[35] J.E. Anderson, J. Phys. Chem. 76 (1972) 4006. Bull. Am. Phys. Soc., 18, 320.

[36] S. Krasen, G. Eiseman, in: G. Eiseman (Ed.), Membranes, vol. 2, Marcel Dekker, New York, 1973, p. 277.

[37] C. Kamizarva, W.J. McCutchan, School of Engineering, University of California, 1973. LA Report No. UCLA-Eng. 7370.

[38] W.A. Duval, T. Helfgott, Removal of wastewater organics by reverse osmosis, Water Poll. Cont. Fed. 47 (1975) 57–65.

[39] T.A. Krug, S. McDougall, Preliminary Assessment of a Microfiltration Reverse Osmosis Process for the Treatment of Landfill/Leachate. Zenon Environmental, Inc. Burlington, Ontario, Canada. Proceedings of the 43rd Industrial Waste Conference, 1988. Purdue University, West Lafayette (IN), 1989, pp. 185–193.

[40] E.S.K. Chian, F.B. De Walle, Sanitary landfill leachates and their treatment, Am. Soc. Civil Eng. Environ. Eng. Div. J. 102 (EE2) (1976) 411–431.

[41] H. Whittaker, C.L. Adams, S.A. Salo, A. Morgan, Reverse osmosis at the gloucester landfill, Proceedings of the Technical Seminar on Chemical Spills, Environment Canada, Toronto, 1991. 190–207.

[42] S. Sourirajan, Reverse Osmosis, Academic Press, New York, 1970, p. 580.

[43] M.T. Brunelle, Colloidal fouling of reverse osmosis membranes, Desalination 32 (1980) 127–135.

[44] D.E. Potts, R.C. Ahlert, S.S. Wang, A critical review of fouling of reverse osmosis membranes, Desalination 36 (1981) 238–253.

[45] W. Eykamp, Fouling of membranes in food processing, AIChE Symposium Series 74 (1976) 234.

[46] L. Sun, E.M. Perdue, J.F. McCarthy, Using reverse osmosis to obtain organic matter from surface and ground water, Water Res. 29 (1995) 1471–1477.

[47] R.N. Kinman, J.W. Stamm, D. Nutini, Reverse Osmosis Membrane Fouling by Sanitary Landfill Leachates and Copper Plating Solutions. Proceedings of the 30th Industrial Waste Conference, 1985, Purdue University, West Lafayette (IN), 1985, 467–475.

[48] T.N. Eisenberg, E.J. Middlebrooks, Reverse Osmosis Treatment of Drinking Water, Butterworths, London, 1986, pp. 163–182.

[49] J.J. Allard, M. Rovel, R. Treille, Before RO pretreatment, Water Sewage Works 126 (1979) R-102.

[50] D. Comstock, Improve RO pretreatment, Water Waste Eng. 17 (1980) 49.

[51] D.G. Argo, Water Factory 21 Design and Operating Data. OWRT PB-300602, Office of Water Research and Technology, Washington, DC, 1979, p. 131.

[52] E. Zaganiaris, G.H. Beasley, Deionization of boiler feedwater: silica removal by counter current regenerated anionic systems, in: D. Naden, M. Streat (Eds.), Ion Exchange Technology, Ellis Norwood, London, 1984, pp. 37–49.

[53] H. Cohen, P. Leader, Membrane processes combine with more conventional separation techniques, Pilot Treatment of Industrial Effluents in Israel, Purdue University, West Lafayette (IN), 1973. Proceedings of the 28th Industrial Waste Conference, Part II, 1973, pp. 623–632.

[54] W.A. Duval, T. Helfgott, Removal of wastewater organics by reverse osmosis, Water Poll. Cont. Fed. 47 (1975) 57–65.

[55] J.E. Drewes, M. Reinhard, P. Fox, Comparing microfiltration-reverse osmosis and soil-aquifer treatment for indirect potable reuse of water, Water Res. 37 (15) (2003) 3612–3621.

[56] S.F. Kang, H.Y. Yen, M.H. Yang, Treatment of textile effluents by H2O2/UV oxidation combined with RO separation for reuse, J. Environ. Sci. Health A Tox. Hazard Subst. Environ. Eng. 38 (7) (2003) 1327–1339.

[57] T.K. Chen, J.N. Chen, Combined membrane bioreactor (MBR) and reverse osmosis (RO) system for thin-film transistor-liquid crystal display TFT-LCD, industrial wastewater recycling, Water Sci. Technol. 50 (2) (2004) 99–106.

[58] S. Lee, R.M. Lueptow, Toward a reverse osmosis membrane system for recycling spacemission wastewater, Life Support Biosph. Sci. 7 (3) (2000) 251–261.

[59] J.B. Downing, E. Bracco, F.B. Green, A.Y. Ku, T.J. Lundquist, I.X. Zubieta, W.J. Oswald, Low cost reclamation using the Advanced Integrated Wastewater Pond Systems Technology and reverse osmosis, Water Sci. Technol. 45 (1) (2002) 117–125.

[60] T. Srinivasan, S. Sridhar, M. Ramakrishna, Treatment of synthetic glucose media by an integrated process of anaerobic digestion and reverse osmosis, J. Environ. Sci. Eng. 51 (3) (2009) 199–206.

[61] C. Murray-Gulde, J.E. Heatley, T. Karanfil, J.H. Rodgers Jr., J.E. Myers, Performance of a hybrid reverse osmosis-constructed wetland treatment system for brackish oil field produced water, Water Res. 37 (3) (2003) 705–713.

[62] X. Yu, Z. Zhong, W. Xing, Treatment of vegetable oily wastewater using an integrated micro-filtration-reverse osmosis system, Water Sci. Technol. 61 (2) (2010) 455–462.

[63] E. Steinle-Darling, M. Zedda, M.H. Plumlee, H.F. Ridgway, M.E. Reinhard, valuating the impacts of membrane type, coating, fouling, chemical properties and water chemistry on reverse osmosis rejection of seven nitrosoalklyamines, including NDMA, Water Res. 41 (17) (2007) 3959–3967.

[64] B. Tepus, M. Simonic, I. Petrinić, Comparison between nitrate and pesticide removal from ground water using adsorbents and NF and RO membranes, J. Hazard Mater. 170 (2–3) (2009) 1210–1217.

[65] N. Ben Amar, N. Kechaou, J. Palmeri, A. Deratani, A. Sghaier, Comparison of tertiary treatment by nanofiltration and reverse osmosis for water reuse in denim textile industry, J. Hazard Mater. 170 (1) (2009) 111–117.

[66] S.J. You, J.Y. Teng, Performance and dye-degrading bacteria isolation of a hybrid membrane process, J. Hazard Mater. 172 (1) (2009) 172–179.

[67] F.S. Kegel, B.M. Rietman, A.R. Verliefde, Reverse osmosis followed by activated carbon filtration for efficient removal of organic micropollutants from river bank filtrate, Water Sci. Technol. 61 (10) (2010) 2603–2610.

[68] S.K. Park, J.Y. Hu, Assessment of the extent of bacterial growth in reverse osmosis system for improving drinking water quality, J. Environ. Sci. Health A Tox. Hazard Subst. Environ. Eng. 45 (8) (2010) 968–977.

[69] V. Gabet-Giraud, C. Miège, J.M. Choubert, S.M. Ruel, M. Coquery, Occurrence and removal of estrogens and beta blockers by various processes in wastewater treatment plants, Sci. Total Environ. 408 (19) (2010) 4257–4269.

[70] S. Theepharaksapan, C. Chiemchaisri, W. Chiemchaisri, K. Yamamoto, Removal of pollutants and reduction of bio-toxicity in a full scale chemical coagulation and reverse osmosis leachate treatment system, Bioresour. Technol. 102 (9) (2011) 5381–5388.

[71] M.A. Boleda, M.A. Galceran, F. Ventura, Behavior of pharmaceuticals and drugs of abuse in a drinking water treatment plant (DWTP) using combined conventional and ultrafiltration and reverse osmosis (UF/RO) treatments, Environ. Pollut. 159 (6) (2011) 1584–1591.

[72] Q. Sui, J. Huang, S. Deng, G. Yu, Q. Fan, Occurrence and removal of pharmaceuticals, caffeine and DEET in wastewater treatment plants of Beijing, China, Water Res. 44 (2) (2010) 417–426.

[73] K. Kimura, T. Iwase, S. Kita, Y. Watanabe, Influence of residual organic macromolecules produced in biological wastewater treatment processes on removal of pharmaceuticals by NF/RO membranes, Water Res. 43 (15) (2009) 3751–3758.

[74] L. Di Palma, P. Ferrantelli, C. Merli, E. Petrucci, Treatment of the solution extracted from metal contaminated soils by reverse osmosis and chemical precipitation, Ann. Chim. 93 (12) (2003) 1005–1011.

[75] H. Topp, H. Russell, J. Davidson, D. Jones, V. Levy, M. Gilderdale, S. Davis, R. Ring, G. Conway, P. Macintosh, L. Sertorio, Process water treatment at the Ranger uranium mine, Northern Australia, Water Sci. Technol. 47 (10) (2003) 155–162.

[76] M. Walker, R.L. Seiler, M. Meinert, Effectiveness of household reverse-osmosis systems in a Western U.S. region with high arsenic in groundwater, Sci. Total Environ. 389 (2–3) (2008) 245–252.

[77] I.S. Chang, E.W. Lee, S. Oh, Y. Kim, Comparison of SAR (sodium adsorption ratio) between RO and NF processes for the reclamation of secondary effluent, Water Sci. Technol. 51 (6–7) (2005) 313–318.

[78] E.M. Voorthuizen, A. Zwijnenburg, M. Wessling, Nutrient removal by NF and RO membranes in a decentralized sanitation system, Water Res. 39 (15) (2005) 3657–3667.

[79] G.C. Ganzi, P.L. Parise, The production of pharmaceutical grades of water using continuous deionization post-reverse osmosis, J. Parenter. Sci. Technol. 44 (4) (1990) 231–241.

[80] J.A. López-Ramírez, S. Sahuquillo, D. Sales, J.M. Quiroga, Pre-treatment optimisation studies for secondary effluent reclamation with reverse osmosis, Water Res. 37 (5) (2003) 1177–1184.

[81] P. Xu, J.E. Drewes, C. Bellona, G. Amy, T.U. Kim, M. Adam, T. Heberer, Rejection of emerging organic micropollutants in nanofiltration-reverse osmosis membrane applications, Water Environ. Res. 77 (1) (2005) 40–48.

[82] A.M. Comerton, R.C. Andrews, D.M. Bagley, Evaluation of an MBR-RO system to produce high quality reuse water: microbial control, DBP formation and nitrate, Water Res. 39 (16) (2005) 3982–3990.

[83] O.A. Jones, P.G. Green, N. Voulvoulis, J.N. Lester, Questioning the excessive use of advanced treatment to remove organic micropollutants from wastewater, Environ. Sci. Technol. 41 (14) (2007) 5085–5089.

[84] J.J. Qin, M.H. Oo, M.N. Wai, C.M. Ang, F.S. Wong, H. Lee, A dual membrane UF/RO process for reclamation of spent rinses from a nickel-plating operation- A case study, Water Res. 37 (13) (2003) 3269–3278.

[85] L. Thörneby, W. Hogland, J. Stenis, L. Mathiasson, P. Somogyi, Design of a reverse osmosis plant for leachate treatment aiming for safe disposal, Waste Manag. Res. 21 (5) (2003) 424–435.

[86] H. Hasar, S.A. Unsal, U. Ipek, S. Karatas, O. Cinar, C. Yaman, C. Kinaci, Stripping/flocculation/membrane bioreactor/reverse osmosis treatment of municipal landfill leachate, J. Hazard Mater. 171 (1–3) (2009) 309–317.

[87] C.R. Bartels, M. Wilf, K. Andes, J. Iong, Design considerations for wastewater treatment by reverse osmosis, Water Sci. Technol. 51 (6–7) (2005) 473–482.

[88] C.P. Chu, Y.S. Wu, C.C. Lin, Y.J. Chung, Reuse potential analysis on WWTP effluent of industrial parks in Taiwan, Water Sci. Technol. 55 (1–2) (2007) 425–431.

[89] R.A. Diltz, T.V. Marolla, M.V. Henley, L. Li, Reverse osmosis processing of organic model compounds and fermentation broths, Bioresour. Technol. 98 (3) (2007) 686–695.

[90] R. Ordóñez, D. Hermosilla, I. San Pío, A. Blanco, Replacement of fresh water use by final effluent recovery in a highly optimized 100% recovered paper mill, Water Sci. Technol. 62 (7) (2010) 1694–1703.

5 Water Treatment by Membrane Filtration Techniques

CHAPTER OUTLINE

5.1 Introduction

Among various water treatment technologies, filtration is considered as a good alternative for secondary and tertiary treatment [1]. About 500 commercial membrane bioreactors (MBRs) are in use globally [2]. The main membrane technology works without the addition of chemicals, with a relatively low-energy involvement. These are suitable for treating ground water, surface water and waste water. The membrane separation process is based on the presence of semipermeable membranes. The membrane acts as specific filter that will let water flow through, while it withholds suspended solids and other substances. The substances are forced to pass through the membrane by high pressure, the concentration gradient and electric potential are maintained on both sides of the membrane. Membranes are holed up in selective separation wall and some substances can pass through the membrane, while others are retained. The membrane filtration can be used as alternatives for flocculation, sediment purification techniques, adsorption

Environmental Water, http://dx.doi.org/10.1016/B978-0-444-59399-3.00005-2

(sand filters and active carbon filters, ion exchangers), extraction, and distillation. The selectivity and productivity (membrane dependent) determined the efficiency of a membrane filtration process. Selectivity is expressed as a parameter called retention or separation factor while productivity is expressed as a parameter called flux [3,4].

Although granular media filters can be used for treating a variety of wastewaters, but its incapability of removing 100% pollutants has developed membrane filtration method. This technology is being used for water treatment for last few decades but advanced development has emerged during last 10 years. Membrane filtration includes reverse osmosis (RO), nanofiltration (NF), microfiltration (MF), membrane cartridge filtration (MCF), and ultrafiltration (UF). RO has not been included in this chapter as it has received wide applications and readers can read the specific chapter devoted on this technique. The membrane technology has been described herein in detail. Basically, membrane filtration is defined as a pressure- or vacuum-driven separation process in which particulate matter larger than 1.0 μm is rejected by membranes. The membranes are manufactured as flat sheets or as hollow fibers and configured into membrane modules. The most common types are hollow fiber, spiral wound, and cartridges. The size of membranes is adjusted in such a way so that they can remove viruses, bacteria, and other microbes. Membrane filtration is of micro-, ultra- and nano-types. The use of membrane type depends on applications. For the removal of larger particles, MF and UF are applied, while NF is used for removal of salts. The pressure required for NF is higher than for MF and UF. The main features of membrane technology are its working capabilities at low temperatures and low energy [5–7].

5.2 Features of Membranes

These are characterized by their ability to remove suspended or colloidal particles by a sieving mechanism based on the size of the membrane pores. All the membrane pore sizes are distributed as per membrane material and manufacturing process. MF and UF membranes have pore sizes ranging from 0.1–0.2 to 0.01–0.05 μm, respectively. NF membranes are often used in the removal of dissolved contaminants, as in the case of softening or desalination. NF utilizes semipermeable membranes and does not have definable pores (0.001 μm). NF achieves removal of dissolved contaminants through the process of RO. The principles of MCF are similar to MF/UF in that MCF removes particles by a sieving mechanism. MCF has not traditionally been considered a membrane treatment process, a cartridge filtration device that utilizes membrane filtration media removing particles of 1.0 μm or larger. Besides, many cartridge filters have pore sizes that are larger than 1.0 μm [8,9]. The characteristic features of different membranes are given in Table 5-1 [10]. On the other hands, Table 5-2 [11] depicts the properties of some membranes.

5.3 Types of Membranes

Tubular membranes have diameter of about 5–15 and are located on the inside of a tube made of different materials. Basically, these materials are supporting media for the

Table 5-1 The Membrane Features for Water Treatment

Membrane Materials	Pore Size	Configurations	Refs.
Polyethylene	0.1 μm	Hollow fiber (dead end)	[67]
Polysulfone	10 kDa	Plate and frame (cross-flow)	[68]
Polysulfone	10 kDa	Flat sheet (cross-flow)	[68]
Polypropylene	0.2 μm	Hollow fiber (dead end)	[69]
Polysulfone	40 kDa	Capillary (cross-flow)	[70]
Polypiperazine	290 Da	Flat sheet (cross-flow)	[71]
Polypropylene	0.2 μm	Hollow fiber (dead end)	[72]
Polyamide TFC	8 kDa	Flat sheet (cross-flow)	[73]
Polypiperazine	360 g/mol	Flat sheet (spiral wound)	[74]
Polyethylene	0.1 μm	Hollow fiber (submerged)	[75]
Polypropylene	0.2 μm	Flat sheet (dead end)	[76]
Polysulfone	0.01 μm	Hollow fiber (cross-flow)	[77]
Composite (PA + PS)	0.65 nm	Flat sheet (cross-flow)	[78]
PAN	50 kDa	Hollow fiber (cross-flow)	[79]
Polysulfone	100 kDa	Hollow fiber (cross-flow)	[80]
PES	2000 g/mol	Flat sheet (spiral wound)	[74]
Metal	0.2 μm	Plate and frame (submerged)	[81]
PAN	110 kDa	Flat sheet (cross-flow)	[82]
Regenerated cellulose	3 kDa	Flat sheet (cross-flow)	[73]
PVDF	0.22 μm	Flat sheet (dead end)	[83]
Sulfonated PES	20 kDa	Flat sheet (cross-flow)	[84]
PES	0.16 μm	Flat sheet (dead end)	[85]
Acrylic	50 kDa	Hollow fiber (cross-flow)	[86]
PVDF	0.22 μm	Flat sheet (dead end)	[87]
Cellulose ester	0.22 μm	Flat sheet (dead end)	[87]
PES	100 kDa	Flat sheet (dead end)	[87]
Cellulosic	100 kDa	Hollow fiber (cross-flow)	[86]
Regenerated cellulose	100 kDa	Flat sheet (dead end)	[88]
Sulfonated PES	0.7 nm	Flat sheet (cross-flow)	[88,89]

membranes. The flow in a tubular membrane is usually inside out due to inner location of tubular system. It is due to weak attachment of the membrane to the supporting media. The plugging of tubular membranes is rare owing to the size of the membrane surface. The drawback of tubular membranes includes low packing density. The diameter of capillary membranes ranged from 0.5 to 5 mm, which is much smaller than that of

Table 5-2 Properties of Cross-flow Methods

Methods	Molecular Weight	Pore Size (nm)	Pressure (bar)	Permeation
MF	>500 kDa	50–5000	0.5–2	As above plus colloids
UF	2–500 kDa	5–50	0.5–10	As above plus macromolecules
NF	500–2000 Da	0.6–5	10–40	Water, low molecular solutes
RO	<500	<0.6	30–70	Water

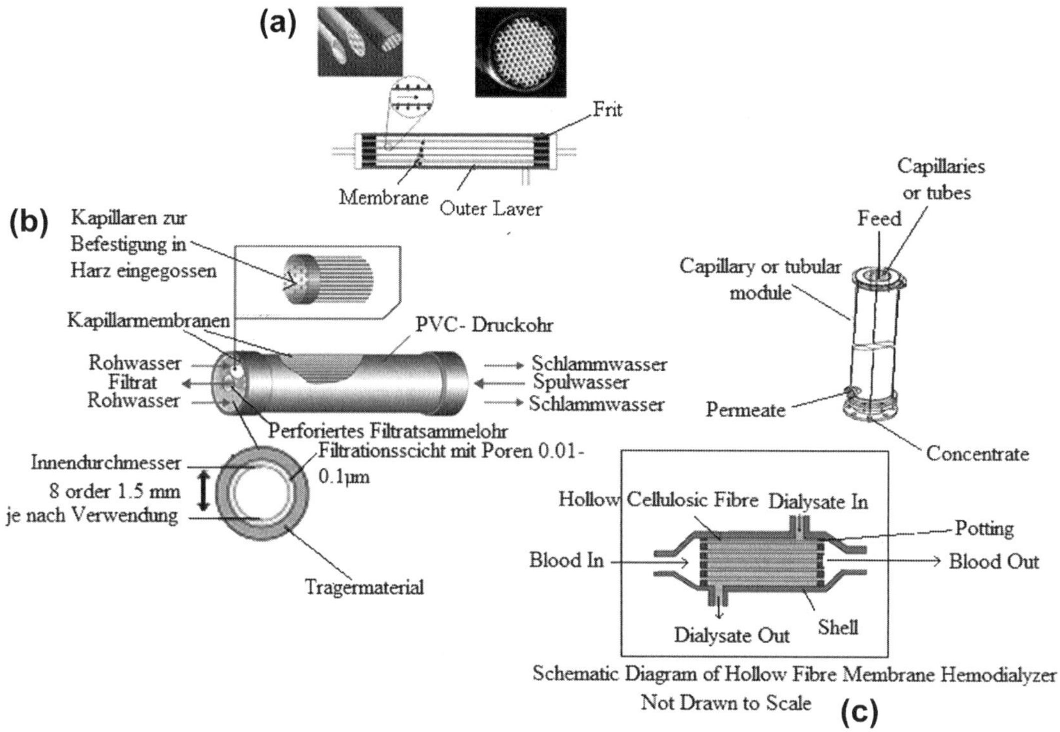

FIGURE 5-1 The different shapes and structures of some membranes: (a) tubular, (b) capillary, and (c) hollow-fiber membranes.

tubular membranes. These membranes work as a selective barrier, which is sufficiently strong to resist filtration pressures. The flow through capillary membranes can be both inside out and outside in. There are many chances of plugging due to smaller diameters. Hollow-fiber membranes are membranes have diameter of about 0.1 μm and, hence, the chances of plugging are very high. These are useful for treating water of low suspended solids. The packing density of a hollow-fiber membrane is very high. Spiral membranes are only used for nano and two layers of membrane are placed onto a permeate collector fabric. These are wrapped around a centrally placed permeate drain. It leads the packing density high and feed channel is placed at height, to prevent plugging of the membrane unit. Pillow-shaped membranes have flat plates in the form of pillow consisting supporting plates. The multiple pillows are placed with a fixed distance depending on the type of water being treated [12,13]. Figure 5-1 indicates the shapes and structures of some membranes. Besides, the process schematic of the two-stage UF plant and NF is also shown in Figures 5-2 and 5-3 [14]. On the other hands, Figure 5-4 shows the filtration process with the application of capillary membranes. Figure 5-5 is [15] an overview of the relevant dimensions in membrane filtration. The pores of filtration membranes range

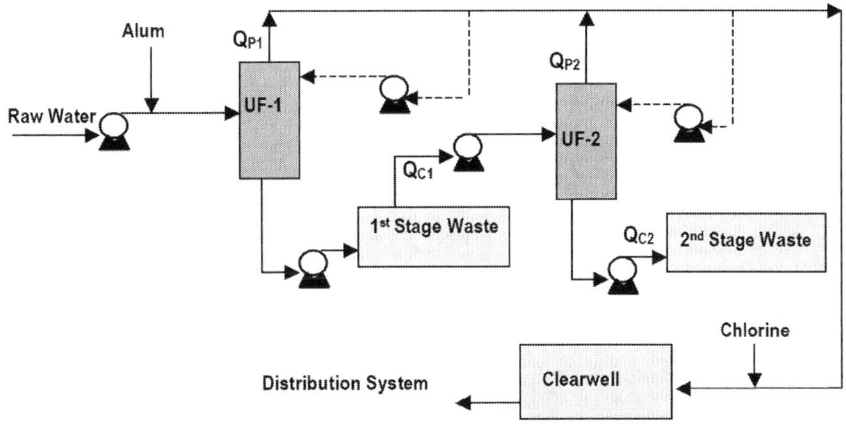

FIGURE 5-2 The process schematic of the two-stage UF plant.

from about 0.5 to more than 1000 nm. The operating strategies in foam ball cleaning system are shown in Figure 5-6 [15].

5.4 Membrane Filtration, Fouling, and Cleaning

The filtration properties of a membrane depend on the pore size and formation of a cake layer during operation. Membrane filtration systems can be optimized by dead end flow and cross-flow to achieve the highest production for a long time. The selection of membrane depends on costs, risks of plugging of the membranes, packing density, and cleaning opportunities. Membranes are designed in several modules such as tubular, plate, and frame membrane systems. Tubular membrane systems are of tubular, capillary, and hollow-fiber types, while plate and frame are divided into spiral- and pillow-shaped structures. Membrane fouling is a normal phenomenon during water treatment

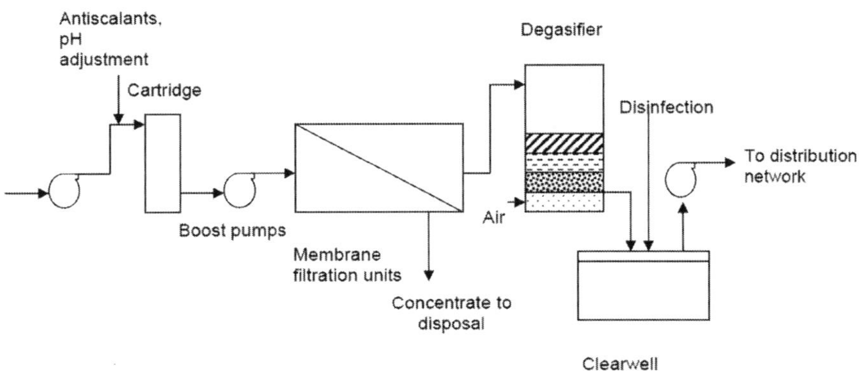

FIGURE 5-3 Typical NF softening plant.

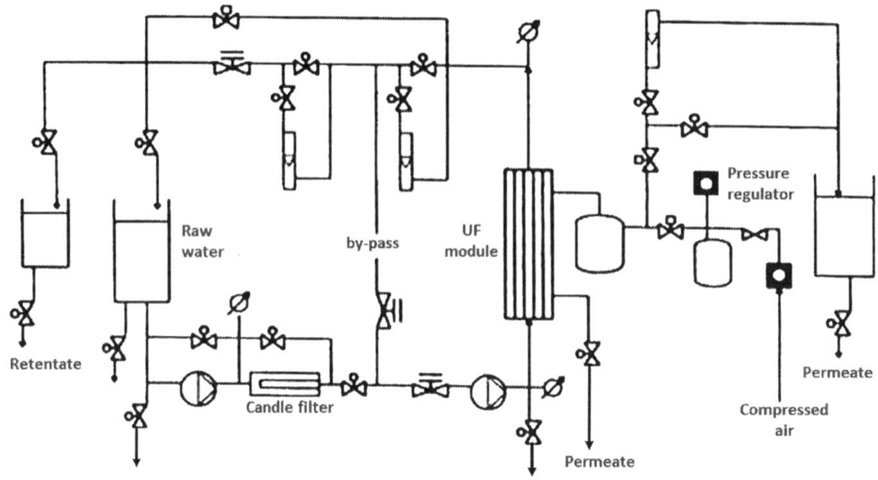

FIGURE 5-4 The filtration process with the application of capillary membranes. For color version of this figure, the reader is referred to the online version of this book.

even with a sufficient pretreatment [16–21]. These depend on feed water quality, membrane type, membrane materials, process design, and control. Basically, main fouling are particles, biofouling, and scaling and cause high workload. There are different cleaning methods such as flushing, backward flushing, air flushing, chemical cleaning, and any combination are the effective methods for cleaning membranes. In forward, flush membranes are flushed with feed water or permeate forward. The particles that are absorbed to the membrane are released and discharged because of the more rapid flow and the resulting turbulence. But, sometimes, these particles can only be removed by backward flushing. In backward, flush permeate is flushed through feed waterside of under pressure, applying twice the flux used during filtration. In case of no fully restore, a chemical cleaning process is applied. The chemical cleaning involves the soaking of membrane with a solution of chlorine bleach, hydrochloric acid, or hydrogen peroxide. It has also been observed that after this, a forward or backward flush is useful for complete cleaning. In air flushing, it is injected in the supplier pipe to create more turbulent [22,23].

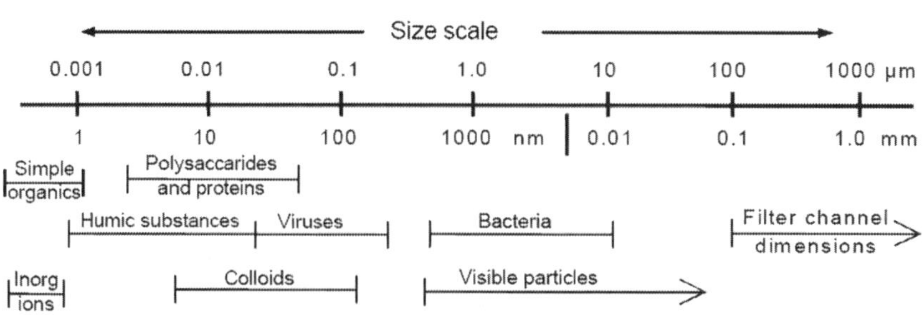

FIGURE 5-5 The relevant dimensions in membrane filtration.

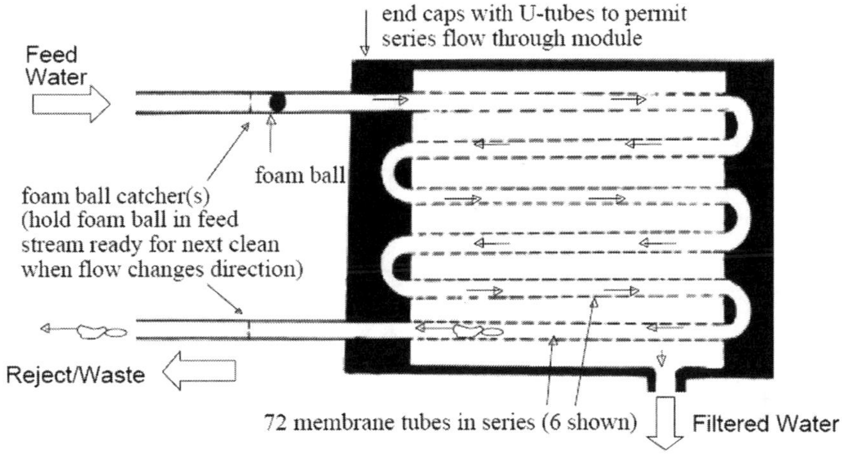

FIGURE 5-6 The operating strategies in foam ball cleaning system.

5.5 Membrane Management

The management of membrane is very important aspect in water treatment for fast and economic method. This job can be achieved via dead end filtration and cross-flow filtration. In dead end filtration, water is pressed through the membrane and stayed on the membrane. It is controlled by pore size of the membrane and feed water pressure decreasing in flux. The pressure that is needed to press water through a membrane is called transmembrane pressure (TMP), pressure gradient of the membrane. In cross-flow filtration, feed water is recycled with water flow parallel to the membrane. Only a small part of the feeds, water is used for permeate production while the major part leaves the module resulting into high-energy cost. Generally, water speed of the feed water is high for controlling the thickness of the cake. This enables the suspended solids to be carried away in the water flow. Basically, this method is useful for NF, UF, and MF, depending on the pore size of the membrane [24,25].

5.6 Membrane Materials

Generally, the membranes are manufactured by synthetic polymer but other materials such as ceramic and metals are also used. But former materials are preferred due to inexpensiveness nature. Most commonly used materials are polyvinylidene fluoride (PVDF), polyacrylonitrile (PAN), cellulose acetate (CA), polypropylene (PP), polyethersulfone (PES), polysulfone (PS), or other polymers. These have different properties such as degree of hydrophobicity, surface charge, tolerance, pH and oxidant strength, and flexibility. The nature of materials determines the properties of membranes, i.e. hydrophilic and hydrophobic. The membranes of polymers reacting with oxidants cannot be used with chlorinated water. Besides, material properties affect the exclusion properties of a membrane, especially charged species. The materials of good strength can bear high pressure and allowing good operational flexibility. Cellulose membranes (commonly used in NF) are susceptible to biodegradation at 4–8 pH. The best chlorinated dose for this sort of membrane is 0.5 mg/L. In spite of low limited tolerance, polyamide (PA) membranes can be used under a wide range of pH ranges with compatibility to work with weaker oxidants such as chloramines. These require low pressure and useful tool for NF [26,27].

5.7 Principles of Membrane Filtration System Design and Operation

The knowledge of the basic principles of membrane filtration system design and operation is important for applying the technology to water treatment. In spite of utilizing pressure (or vacuum) as a driving force, all MF, UF, NF, and MCF have differences in the models. Important concepts are flux, recovery, and flow balance. The membrane filtration system is characterized by the system flux and defined as the filtrate flow per unit of membrane filtration area. The following equation can be used for the purpose:

$$J = Q_p/A_m \tag{5.1}$$

where J = flux (gfd), Q_p = filtrate flow (gpd), and A_m = membrane surface area (ft^2).

The recovery of a membrane unit is defined as the amount of feed flow and converted to filtrate flow. It is expressed as a decimal percent and given in the following equation:

$$R = Q_p/Q_f \tag{5.2}$$

where R = recovery of the membrane unit (chemical percentage), Q_p = filtrate flow produced by the membrane unit (gpd), and Q_f = feed flow to the membrane unit (gpd).

A general flow balance that can be applied to all membrane filtration systems is shown in the following equation:

$$Q_f = Q_c + Q_p \tag{5.3}$$

where Q_f = feed flow to the membrane unit (gpd), Q_c = concentrate flow from the membrane unit (gpd), and Q_p = filtrate flow from the membrane unit (gpd).

The concentrate (bleed or reject) flow Q_c is zero for systems operating without a concentrate waste or bleed stream [systems operated in deposition (dead end)] mode or cross-flow systems in which 100% of the concentrate is recirculated. For designing membrane filtration system, it is desirable to account the additional filtered water used for both backwashing and chemical cleaning in the determination of the filtrate flow Q_p. Similarly, an estimate of the total required feed flow Q_f to the system should be incorporated any raw water that might be used in these routine maintenance processes [28,29].

5.8 Applications

The membranes are used for the removal of various pollutants from water. But the specific use depends on the other quality of water and the amount treated. Some specific

Table 5-3 The Applications of Membranes

Membrane Type	Applications
UF	Removes viruses and large macromolecules recovery in water recycling.
MF	Removes turbidity and microorganisms.
NF	Removes color, large ions, and higher flux than RO for surface water treatment.
MBR	Treat difficult waste possibly with specific organism for special waste.
Hybrid	UV–ozone–precoagulation or combined membrane systems becoming much effective than a single membrane.
RO	Potable use and industrial reuse removal of small ions.

applications of different membrane technologies are summarized in Table 5-3 [10]. Some examples of their applications are discussed in the following subsections.

5.8.1 Organic Pollutants

As discussed earlier, organic pollutants are very dangerous and require special applications of membranes. Marcucci et al. [30] applied MF or UF membrane processes for the removal of dyes from textile wastewater. The spiral wound membranes, operating under pressure, to flat membranes, operating under vacuum, were compared. The results and a preliminary economic analysis indicated the possibility of the membrane technologies to an industrial scale for textile wastewater. The authors used the developed application at pilot level. Cheng et al. [31] applied charged NF membrane technology for direct treatment of textile wastewater for removal of cations, dyes, and heavy metals. The optimal process conditions for the removal of small cations using this membrane is in the region pH <8.0 and concentration less than 15 mol M. The authors tested their method first in laboratory by removing methylene blue from a synthetic dye house wastewater. The PA6DT-C membrane was successful in removal of methylene blue dye from synthetic dye house wastewaters achieving 98% rejection and a membrane flux of ≈ 17 LMH/bar. The membrane flux for membrane was superior to the commercial membranes with an approximate increase of threefold to fourfold. Cassano et al. [32] removed phenols from olive mill wastewaters during olive oil production by using UF membranes. The UF membranes were characterized by different molecular weight cutoff (4, 5, and 10 kDa) and polymeric material (regenerated cellulose and PES) and used for the purpose. The authors evaluated permeate fluxes, fouling index, and retention coefficients toward phenolic compounds, total antioxidant activity (TOC) and total organic carbon (TOC).

Ollis [33] reported the removal of organic contaminants by hybrid process. The chemical oxidation photocatalysis process with MF and UF is considered to be the best approach. Wintgens et al. [34] reported the removal of estrogenic pollutants from landfill leachate. The authors used MBRs and NF as the suitable techniques. Besides, xenoestrogenic substances such as nonylphenol and bisphenol have been removed from landfill leachate. Kim et al. [35] used NF90 for removing pharmaceutically active compounds (PhACs) from water. The pollutants removed were pesticides, solvents, and endocrine disrupting compounds. According to authors, the transport of these solutes through these membranes was facilitated by solute membrane interactions. The solute–membrane interactions were estimated by diffusion coefficients through membranes pores. Beier et al. [36] used pretreatment MBR and advanced treatment high-pressure membrane system technologies for the removal of bezafibrate, bisoprolol, carbamazepine, clarithromycin, ciprofloxacin, diclofenac, ibuprofen, metronidazole, moxifloxacin, telmisartan, and tramadol from hospital wastewater in Germany. The results of this study confirmed that MBR technology followed by an advanced treatment for trace pollutant removal was an adequate approach for specific treatment of hot spot wastewater such as hospital wastewater. The schematic diagram of the large scale MBR at the hospital Waldbrol is shown in Figure 5-7. Furthermore, the same authors [37] used MBRs as an attractive option for the treatment of hospital wastewater and elimination of pharmaceuticals in high-density urban areas. As per the authors, the segregation of these wastewaters and their separate treatment at the source could be reduced the entry of drugs in waterways and enabled water reuse after adequate polishing treatment processes. Bruchet and Laîné [38] reported that membrane processes have potential to remove of taste-and-odor-causing compounds. Several membranes were tested including UF, UF combined with powdered activated carbon (PAC), NF, and low-pressure RO (LP RO) membranes. The results of this study indicate that the combination of UF with PAC is effective for T&O control, whereas the benefit of NF and LP RO remains unclear for T&O control. Nataraj et al. [39] used NF to remove the color and contaminants of the distillery spent wash. According to authors, the pollutant level in permeates was

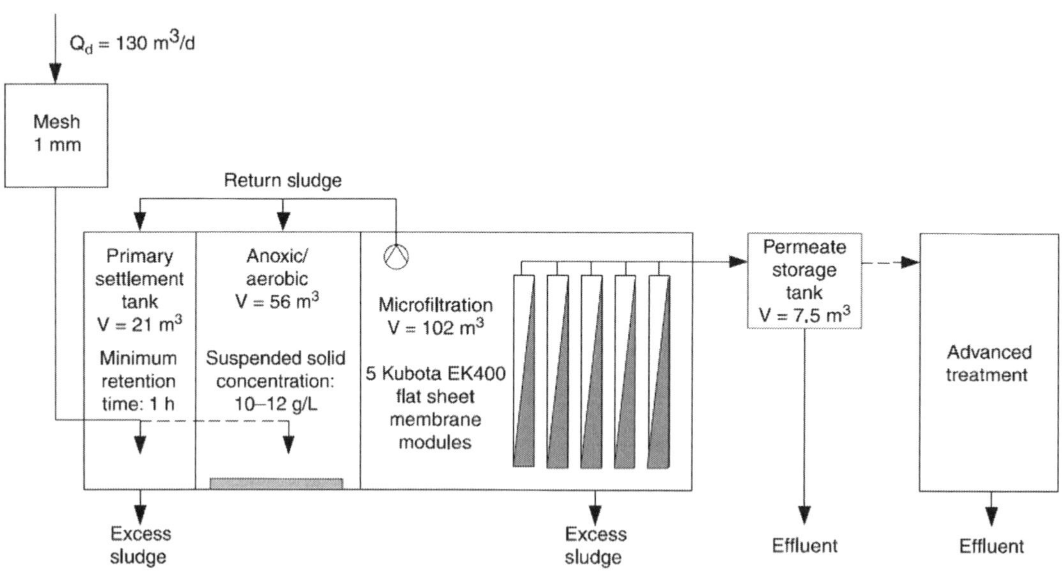

FIGURE 5-7 The schematic diagram of the large-scale MBR at the hospital Waldbrol.

below the maximum contaminant level as per the guidelines of the World Health Organization. Szép and Kohlheb [40] used pretreatments, membrane filtration, and post-treatment methods for treating wastewater of oil and gas production units. This combination decreased conductivity by 98%, chemical oxygen demand (COD) by 100%, and the SAR by 2.15 mgeqv (0.5) in the produced water. According to the authors, the cost of produced water was less expensive than to dispose of it by injection and this treated water might be of great value in water-poor regions. The authors concluded the mobile station as provided a viable and cost-effective result beneficial for use of produced water.

Fakhru'l-Razi et al. [41] reported the performance of a membrane sequencing batch reactor (MSBR) for treating water oil and gas-field wastewater. The MSBR was operated in different hydraulic residence time (HRT) of 8, 20, and 44 h. The results showed that for a HRT of 20 h, the combined process effluent COD, TOC, and oil and grease (O&G) removal efficiencies were 90.9%, 92%, and 91.5%, respectively. According to the authors, foulant biopsy demonstrated that the fouling on the membrane surface was mainly due to inorganic (salts), organic (hydrocarbons), and biological (microorganisms) contaminants. Salahi et al. [42] reported the purification of oily wastewater by UF process. Laboratory-scale UF using polysulfone (PS) and polyacrylonitrile (PAN) membranes were employed with typical oily wastewater collected from API unit of Tehran refinery. The PAN membrane showed higher rejection, more permeation flux, and less fouling resistance than the PS membrane. Both membranes produced permeate with O&G contents generally less than 5 ppm. Rejection of COD and biological oxygen demand (BOD5) were found to be 65% for UF treatment. For further treatment of the UF permeates, Gürel and Büyükgüngör [43] used MBR for removal of organic substances and nutrients from slaughterhouse plant wastewater. The COD, total nitrogen (TN), and total phosphorus (TP) concentrations of slaughterhouse wastewater were found to be 571 mg O_2/L, 102.5 mg/L, and 16.25 mg PO^{4-} P/L, respectively. The removal efficiencies for COD, TOC, TP, and TN were found to be 97%, 96%, 65%, and 44%, respectively. The COD value of wastewater was decreased to 16 mg/L TOC to 9 mg/L. Ammonium and nitrate nitrogen concentrations of treated effluent were 0.100 mg NH^{4+} N/L and 80.521 mg NO^{3-} N/L, respectively. As per the authors, the slaughterhouse wastewater was successfully treated with the MBR process. Chen et al. [44] used a facile in situ approach for constructing tunable amphiphilic or hydrophilic antifouling membranes and used for treating oily water. The membrane with amphiphilic surface showed about 99.5% retention ratio of COD without appreciable membrane fouling. The water permeation flux slightly decreased during operation (total flux decline was 6.8%) and almost completely recovered to the initial value (flux recovery ratio was more than 99.0%) after simple hydraulic washing. For the proteins containing wastewater, the membrane with hydrophilic surface resulted into 52.6% COD retention ratio and superior antifouling performance. Besides, mere 17.0% total flux declined with more than 99.0% flux recovery ratio. The authors hoped that this approach could be developed as a competitive platform technology for the preparation of robust and versatile antifouling membrane, resulting to high process efficiency of wastewater treatments.

Klüpfel et al. [45] used three NF membranes with a chlorine tolerance > or = 1 mg/L to reduce disinfection by-product and their precursors in swimming pool water. The authors installed a lab-scale plant with cross-flow modules at swimming pool for a period of several weeks. The chlorine tolerances of the membranes SB90 and NP030 were found to be enough for filtration under swimming pool water conditions. The retention of dissolved organic carbon (DOC) and adsorbable organic halogens (AOXs) was about 70% and 80% for SB90 and 50% and 40% for NP030, respectively. The results of DOC mass balances for different UF–NF treatment processes showed that pool water quality improved as compared to the conventional treatment system. Linlin et al. [46] reported the removal of dissolved organic matter in municipal effluent by combination of ozonation, slow sand filtration, and NF methods. The main pollutants removed were humic acid and fulvic acid type compounds. DOC and trihalomethane formation potential (THMFP) could be reduced from 6.5 ± 1.1 to 0.7 ± 0.3 mg/L and from 267 ± 24 to 52 ± 6 μg/L, respectively. The very low DOC concentration of 0.6 ± 0.2 mg/L and THMFP of 44 ± 4 μg/L could be reached after the aquifer treatment.

Grilli et al. [47] evaluated the treatability of textile wastewaters in a bench-scale experimental system, comprising an anaerobic biofilter, an anoxic reactor, and an aerobic MBR. The system was effective in the treatment of the textile wastewater by removing COD about 90–95% and color 70%. Beside, the addition of NF membrane allowed further improvement in COD (50–80%), color (70–90%), and salt removal (60–70% as conductivity). Briefly, in NF treatment allowed almost complete removal of the residual color and a reduction of the conductivity. Benito-Alcázar et al. [48] described the treatment of petrochemical wastewater by NF. Although several methods were used for the pretreatment of the petrochemical secondary effluents but UF permeate obtained was suitable for NF and granulated activated carbon filtration coupled with NF provided a better effluent quality of water. Joss et al. [49] reported that high water quality with a water yield of 90% can be achieved by combining a MBR with a submerged UF membrane. The treatment process involves conditioning of MBR effluent prior to the RO and then recycling the RO concentrate.

5.8.2 Inorganic Pollutants

Inorganic pollutants are also notorious and membranes may be used for their removal in water. Katsou et al. [50] used UF or MF membranes for removing lead metal ion from industrial wastewater. The optimized parameters were solution pH, membrane pore size, mineral type, and concentration and mineral–metal contact time. UF system removed high percentage of lead. The combined application of UF/MF membranes effectively removed lead from wastewater resulting in a final effluent. Katsou et al. [51] described the removal of heavy metals from wastewater by using MBR. The average removal efficiencies by the MBR system were 80% for Cu(II), 98% for Pb(II), 50% for Ni(II), and 77% for Zn(II), respectively. Furthermore, the addition of 5 g/L vermiculite into the biological reactor enhanced metal removal to 88% for copper, 85% for zinc, and 60% for nickel, while it reduced biomass inhibition and increased biomass growth. It was also observed that metal ions remaining in soluble form penetrated into the permeate, while those attached to sludge flocs retained by the UF membranes. Furthermore, the average heterotrophic biomass inhibition was 50%, which reduced to 29% at low metal concentrations in the presence of vermiculite. The respective autotrophic biomass inhibition was 70% and 36%. The presence of heavy metals and vermiculite in the mixed liquor adversely impacted on membrane fouling. Furthermore, the same group [52] described the removal of Pb(II), Cu(II), Zn(II), and Ni(II) from industrial wastewater by using sorbent-assisted UF. At pH 6.0, the metal removal sequence by UF system was Pb(II) > Cu(II) > Zn(II) > Ni(II) in milligrams per gram with significant amount of lead and copper being removed due to chemical precipitation and formation of precipitates/complexes with wastewater compounds. While at this pH, zinc and nickel adsorption onto minerals was significant, particularly when bentonite and vermiculite were employed as adsorbents. According to the authors, metal adsorption onto zeolite and bentonite followed the sequence Zn(II) > Ni(II) > Cu(II) > Pb(II), while for vermiculite, the sequence was Ni(II) > Zn(II) > Cu(II) > Pb(II) in milligrams per gram. Low amount of Pb(II) and Cu(II) adsorbed by minerals was attributed to low available lead and copper concentrations. Contrarily, at 9.0 pH, the adoption of UF could effectively reduce heavy metals to very low levels. The similar results were observed at pH 8.0, provided that minerals were added.

Ritchie and Bhattacharyya [53] used MF for removal of sulfate and nitrate. Furthermore, the authors reported MF membrane-based sorbents containing multiple polymeric functional groups effective for the removal of nontoxic metals, such as calcium and magnesium. They claimed these membranes useful with the attachment of various polyamino acids (MW: 2500–10,000), such as polyaspartic acid (cation sorption), polyarginine (oxyanion sorption), and polycysteine (chelation exchange), directly on the membrane pore surfaces. Kurama et al. [54] used NF for the removal of ammonia, nitrite, and calcium. The authors proposed their method for pilot plant application. Koyuncu and Yazgan [55] reported that NF membranes were the best for the treatment of salty and polluted Kucukcekmece Lake. The treated water was used for drinking purpose at Istanbul City. A thin film composite type of spiral wound membrane was used. Myung et al. [56] reported dyeing wastewater pretreatment by using coagulation, activated sludge process, and MF process which was then treated by using NF. It is demonstrated that the extent of fouling was significantly influenced by the surface roughness and the surface charge on the NF membranes. A membrane with a smooth and neutral surface was fouled less. The pretreatment was essential for avoiding NF membranes fouling. The quality of the final permeate was acceptable for water reuse. Leo et al. [57] removed phosphorus from wastewater by using NF membranes. The commercial membranes (DK5, MPF34, NF90, NF270, NF200) were used at pulp and paper plant. As per the authors, NF90 membranes offered the highest rejection of phosphorus of more than 70% for a feed containing 2.5 g/L of phosphorus at a pH <2. But the separation performance of NF90 was slightly affected by phosphorus concentration and pressure, which might be due to concentration polarization and fouling. According to the authors, the separation performance of NF90 was improved in removing phosphorus by adjusting pH to 2 or adding sulfuric acid.

5.8.3 Biological Pollutants

The membranes are used for good removal of microbes and other biological pollutants. Rautenbach et al. [58] combined precipitation/UF processes and used for the removal of bacteria and viruses. The authors developed a new UF module in close cooperation with the Rochem company, Germany, for making the process economic. The plant is in use at municipal wastewater treatment plant (WWTP) Berlin Ruhleben and supposed to be effective system. Xiao-yan et al. [59] used a laboratory MBR having polyethylene hollow-fiber membrane module with a pore size of 0.4 μm and a total surface area of 0.2 m^2 for treatment of raw water is slightly contaminated by domestic sewage. According to the authors, the MBR worked for more than 500 days, with a hydraulic retention time (HRT) being as short as 1 h or less. The method removed turbidity from 4.50 ± 1.11 to 0.08 ± 0.03 NTU, in total coliforms from 105/mL to less than 5/mL and in UV254 absorbance from 0.098 ± 0.019 to 0.036 ± 0.007/cm. The initial chlorine demand for disinfection decreased from 22.3 ± 5.1 to 0.5 ± 0.1 mg/L. The authors claimed their method, i.e. the MBR with a short HRT as an effective biological water treatment process to address the urgent need of many developing countries.

Ruel et al. [60] removed micropollutants at the microgram per liter range in order to meet reuse applications and contribute to reach the good status of the water bodies through seven advanced treatment lines (one MBR process and six tertiary treatment lines). The MBR process allowed to upgrade the removal efficiencies of about 20% of the substances measured. Simmons et al. [61] studied the removal of human entero virus (EV) and noro virus genogroup II (NoV GGII) in a full-scale MBR WWTP. Furthermore, it was compared with the removal of human adenovirus (HAdV). As per the authors, full-scale MBR treatment was able to reduce the viral loads by approximately 5.1 and 3.9 log units for EV and NoV GGII as compared to 5.5 log units for HAdV. This full-scale MBR system outperformed the removal in previous pilot and bench-scale studies by 1–2 log units. To the best of the author's knowledge, it was the first study focusing on the removal of EV in a full-scale MBR WWTP using real-time RT-PCR and on the solid–liquid distribution of EV and NoV GII in secondary biological treatment.

5.8.4 Direct Applications

Thousands of membrane plants are being used worldwide by commercial companies. But the following papers describe the direct applications of membrane units from the research point of views. Laera et al. [62] described sequencing batch MBRs as good option in upgrading small municipal plant and for industrial applications. The authors studied the effects of volumetric exchange ratio and aeration/filtration strategy along with the opportunity of further simplification of the membrane operation by choosing a continuous filtration mode, instead of the usual short cycle of permeation/relaxation. Mondal et al.

[63] used NF for treating wastewater of oil and gas units at Colorado, USA. As per the authors, NF may be a viable process for water treatment. Senta et al. [64] developed MBRs and NF combination for removing many pharmaceuticals from the municipal and industrial wastewater. With further ozonation, various microbes were also removed. The treatment system was found to be effective for the removal of sulfonamides (SAs), fluoroquinolones (FQs), macrolides (MACs), and trimethoprim (TMP). It was observed that the removal of SA in MBR treatment was very efficient, while the elimination of MAC, FQ, and TMP were incomplete.

Sperlich et al. [65] described the reduction of scaling due to phosphate in NF membranes. The chemical precipitation by dosing of sodium hydroxide solution was rapid and removed more than 90% of phosphate and calcium ions. By the removal of calcium ions, chemical precipitation can significantly reduce the scaling potential of NF concentrates, resulting into higher recoveries in NF process. Owen et al. [66] described economic statutes of membrane water treatment. The results of extensive pilot plant trials have been used to determine the cost of treatment using a range of UF and MF membranes. The important factors controlling the cost were membrane cost, membrane replacement frequency, and power. The importance of selecting the suitable membrane and operating conditions for each application was demonstrated. A comparison has been made between the costs of membrane systems and conventional treatment processes.

5.9 Future Perspectives

Membrane technology for water treatment is supposed to be the most promising as it provides good quality of treated water. It is more attractive and popular when the unit is a compact system. Nowadays, the advancement in technology overcomes some membrane limitations such as lowlife, expensiveness, fouling, and permeate quality. Still there are many technical challenges to optimize and make membrane technology more competitive, in the market, large-scale industries and communities. There is a need for further research on advanced membrane materials, which may be resistant to both chemical and mechanical barriers. These may be helpful in prolonging the membrane life span and induce long-term performance. Besides, the optimization of design, treatment configuration, and operating parameters will also be effective and economic in water treatment. The development of fundamental knowledge on understanding and minimization of membrane is also important and can be realized through accurate process design and operating conditions. The proper selection of pretreatments, improvement of cleaning strategies, and membrane systems with low-energy requirements may be helpful to gain the confidence of the market.

Conclusions

Keeping all these facts into consideration, it is important to mention here that membrane filtrations are important water treatment systems depending on the type of water pollutants and other qualities of water. Furthermore, there is a great need to develop effective and efficient membranes for the treatment of wastewater at the commercial scale. The membranes should be prepared with cost-effective material having good permeate flux, should work in the entire pH range, pressure, and temperature, should be resistant to fouling-clogging, should be resistant to chemical and biological attack, and should be versatile in nature. The costly pretreatment of the wastewater should be avoided by optimizing the various parameters such as pH, pressure, nature of membrane, etc. Besides, some inexpensive methods for pretreatment of wastewaters may be developed based on the nature of membranes and the other operating parameters. In a nutshell, there is a great need to develop more inexpensive, fast, and efficient membrane systems for the treatment of leachates and wastewaters of different industries.

References

[1] K.N. Bourgeous, J.L. Darby, T. Tchobanoglous, Ultrafiltration of wastewater: effects of particles, mode of operation, and backwash effectiveness, Water Res. 35 (2001) 77–90.

[2] I.-S. Chang, S.-N. Kim, Wastewater treatment using membrane filtration – effect of biosolids concentration on cake resistance, Process Biochem. 40 (2005) 1307–1314.

[3] D. Livingston, H. Trivedi, Treating the Cause Not the Symptom, Membrane Technology, Supplement to Water & Wastes Digest, Spring, 2006, pp. 18–19.

[4] C.V. Smith, D.O. Di Gregorio, R.M. Talcott, The use of ultrafiltration membranes for activated sludge separation, Proceedings of the 24th Industrial Waste Conference, Purdue University, Lafayette, Indiana, 1969, pp. 805–813.

[5] U.S. EPA, Membrane Filtration Guidance Manual, Proposal Draft (2003), EPA 815-D-03-008.

[6] E.R. Hall, P.R. Bérubé, P.M. Sutton, Membrane Bioreactors for Anaerobic Treatment of Wastewaters. Project 02-CTS-4, Water Environment Research Foundation, 2006.

[7] M.D. Knoblock, P.M. Sutton, P.N. Mishra, K. Gupta, A. Janson, Membrane biological reactor system for treatment of oily wastewaters, Water Environ. Res. 66 (2) (1994) 133–139.

[8] Y. Kiso, Y. Sugiura, T. Kiato, K. Nishimira, Effect of hydrophobicity and molecular size on rejection of aromatic pesticides with nanofiltration membranes, J. Membr. Sci. 192 (2001) (2001) 1–10.

[9] J. Daugherty, K. Alexander, D. Cutler, M. Patel, S. Deshmukh, Applying advanced membrane technology for Orange County's water reuse treatment facilities, Proceedings of the AWWA Membrane Technology Conference, Phoenix, AZ, USA, 2005.

[10] A.W. Zularisam, A.F. Ismail, R. Salim, Behaviours of natural organic matter in membrane filtration for surface water treatment – a review, Desalination 194 (2006) 211–231.

[11] T. Thorsen, Fundamental Studies on Membrane Filtration of Coloured Surface Water, Ph.D. Thesis, NTNU, Trondheim, November 1999.

[12] B.C. Abi-Samra, P. Cook, P. Lange, J. Zepeda, Design and operation of a microfiltration and reverse osmosis system treating secondary effluent and providing high quality reclaimed \water, Proceedings of the AMTA Conference, Tampa, FL, USA, 2002.

[13] T. Walker, C. Stewart, M. Brewster, P. Longfield, Industrial waste organic fouling on membrane recycling plants on Australian experience, Proceedings of the Ozwater Conference, Australia, 2007.

[14] R.A. Bergman, Membrane softening versus lime softening in Florida: a cost comparison update, Desalination 102 (1995) 11–24.

[15] T. Thorsen, H. Fløgstad, Nanofiltration in Drinking Water Treatment, Techneau, 11 December 2006.

[16] L. Defrance, M.Y. Jaffrin, Comparison between filtrations at fixed transmembrane pressure and fixed permeate flux: application to a membrane bioreactor used for wastewater treatment, J. Membr. Sci. 152 (1999) 203–210.

[17] S. Ognier, C. Wisniewski, A. Grasmik, Biofouling in membrane bioreactors: phenomenon analysis and modeling, Proceeding of MBR3 Conference, Cranfield, 16 May 2001, pp. 29–43.

[18] E. Tardieu, A. Grasmick, V. Geaugey, J. Manem, Hydrodynamic control of bioparticle deposition in a MBR applied to wastewater treatment, J. Membr. Sci. 147 (1998) 1–12.

[19] I.S. Chang, P. Le-Clech, B. Jefferson, S.J. Judd, Membrane fouling in membrane bioreactors for wastewater treatment, J. Environ. Eng. 128 (11) (2002) 1018–1029.

[20] I.S. Chang, C.H. Lee, Membrane filtration characteristics in membrane coupled activated sludge system – the effect of physiological states of activated sludge on membrane fouling, Desalination 120 (3) (1998) 221–233.

[21] I.S. Chang, J.S. Kim, C.H. Lee, The effects of EPS on membrane fouling in a MBR process, in: Proceeding of MBR3 Conference, Cranfield, 16 May 2001, pp. 19–28.

[22] R.W. Baker, Membrane Technology and Applications, John Wiley & Sons Ltd, England, 2004.

[23] H. Choi, K. Zhang, D.D. Dionysiou, D.B. Oerther, G.A. Sorial, Effect of permeate flux and tangential flow on membrane fouling for wastewater treatment, Sep. Purif. Technol. 45 (2005) 68–78.

[24] M.J. Hammer, Water and Wastewater Technology, John Wiley & Sons, 1975, ISBN 0-471-34726-4, pp. 390–391.

[25] Metcalf & Eddy, Inc., Wastewater Engineering, McGraw-Hill Book Company, 1972, ISBN 0-07-041675-3, pp. 552–554.

[26] M.T.M. Pendergast, E.M.V. Hoek, A review of water treatment membrane nanotechnologies, Energy Environ. Sci. 4 (2011) 1946–1971.

[27] M. Cheryan, Ultrafiltration and Microfiltration Handbook, Technomic Publishing Co., Ltd, Lancaster, PA, 1998.

[28] S. Judd, The MBR Book. Principles and Applications of Membrane Bioreactors in Water and Wastewater Treatment, Elsevier, Oxford, 2006.

[29] S. Atkinson, Research studies predict strong growth for MBR markets, Membr. Technol. (2006) 8–10.

[30] M. Marcucci, I. Ciabatti, A. Matteucci, G. Vernaglione, Membrane technologies applied to textile wastewater treatment, Ann. N. Y. Acad. Sci. 984 (2003 March) 53–64.

[31] S. Cheng, D.L. Oatley, P.M. Williams, C.J. Wright, Characterisation and application of a novel positively charged nanofiltration membrane for the treatment of textile industry wastewaters, Water Res. 46 (1) (2012 January 1) 33–42.

[32] A. Cassano, C. Conidi, E. Drioli, Comparison of the performance of UF membranes in olive mill wastewaters treatment, Water Res. 45 (10) (2011 May) 3197–3204.

[33] D.F. Ollis, Integrating photocatalysis and membrane technologies for water treatment, Ann. N. Y. Acad. Sci. 984 (2003 March) 65–84.

[34] T. Wintgens, M. Gallenkemper, T. Melin, Occurrence and removal of endocrine disrupters in landfill leachate treatment plants, Water Sci. Technol. 48 (3) (2003) 127–134.

[35] T.U. Kim, G. Amy, J.E. Drewes, Rejection of trace organic compounds by high-pressure membranes, Water Sci. Technol. 51 (6–7) (2005) 335–344.

[36] S. Beier, S. Köster, K. Veltmann, H. Schröder, J. Pinnekamp, Treatment of hospital wastewater effluent by nanofiltration and reverse osmosis, Water Sci. Technol. 61 (7) (2010) 1691–1698.

[37] S. Beier, C. Cramer, S. Köster, C. Mauer, L. Palmowski, H.F. Schröder, J. Pinnekamp, Full scale membrane bioreactor treatment of hospital wastewater as forerunner for hot-spot wastewater treatment solutions in high density urban areas, Water Sci. Technol. 63 (1) (2011) 66–71.

[38] A. Bruchet, J.M. Laîné, Efficiency of membrane processes for taste and odor removal, Water Sci. Technol. 51 (6–7) (2005) 257–265.

[39] S.K. Nataraj, K.M. Hosamani, T.M. Aminabhavi, Distillery wastewater treatment by the membrane-based nanofiltration and reverse osmosis processes, Water Res. 40 (12) (2006 July) 2349–2356.

[40] A. Szép, R. Kohlheb, Water treatment technology for produced water, Water Sci. Technol. 62 (10) (2010) 2372–2380.

[41] A. Fakhru'l-Razi, A. Pendashteh, Z.Z. Abidin, L.C. Abdullah, D.R. Biak, S.S. Madaeni, Application of membrane-coupled sequencing batch reactor for oilfield produced water recycle and beneficial re-use, Bioresour. Technol. 101 (18) (2010 September) 6942–6949.

[42] A. Salahi, T. Mohammadi, Experimental investigation of oily wastewater treatment using combined membrane systems, Water Sci. Technol. 62 (2) (2010) 245–255.

[43] L. Gürel, H. Büyükgüngör, Treatment of slaughterhouse plant wastewater by using a membrane bioreactor, Water Sci. Technol. 64 (1) (2011) 214–219.

[44] W. Chen, Y. Su, J. Peng, X. Zhao, Z. Jiang, Y. Dong, Y. Zhang, Y. Liang, J. Liu, Efficient wastewater treatment by membranes through constructing tunable antifouling membrane surfaces, Environ. Sci. Technol. 45 (15) (2011 August 1) 6545–6552.

[45] A.M. Klüpfel, T. Glauner, C. Zwiener, F.H. Frimmel, Nanofiltration for enhanced removal of disinfection by-product (DBP) precursors in swimming pool water-retention and water quality estimation, Water Sci. Technol. 63 (8) (2011) 1716–1725.

[46] W. Linlin, Z. Xuan, Z. Meng, Removal of dissolved organic matter in municipal effluent with ozonation, slow sand filtration and nanofiltration as high quality pre-treatment option for artificial groundwater recharge, Chemosphere 83 (5) (2011 April) 693–699.

[47] S. Grilli, D. Piscitelli, D. Mattioli, S. Casu, A. Spagni, Textile wastewater treatment in a bench-scale anaerobic-biofilm anoxic-aerobic membrane bioreactor combined with nanofiltration, J. Environ. Sci. Health A Tox. Hazard. Subst. Environ. Eng. 46 (13) (2011 November) 1512–1518.

[48] C. Benito-Alcázar, M.C. Vincent-Vela, J.M. Gozálvez-Zafrilla, J. Lora-García, Study of different pretreatments for reverse osmosis reclamation of a petrochemical secondary effluent, J. Hazard. Mater. 178 (1–3) (2010 June 15) 883–889.

[49] A. Joss, C. Baenninger, P. Foa, S. Koepke, M. Krauss, C.S. McArdell, K. Rottermann, Y. Wei, A. Zapata, H. Siegrist, Water reuse: >90% water yield in MBR/RO through concentrate recycling and CO_2 addition as scaling control, Water Res. 45 (18) (2011 November 15) 6141–6151.

[50] E. Katsou, S. Malamis, K. Haralambous, Pre-treatment of industrial wastewater polluted with lead using adsorbents and ultrafiltration or microfiltration membranes, Water Environ. Res. 83 (4) (2011 April) 298–312.

[51] E. Katsou, S. Malamis, M. Loizidou, Performance of a membrane bioreactor used for the treatment of wastewater contaminated with heavy metals, Bioresour. Technol. 102 (6) (2011 March) 4325–4332.

[52] E. Katsou, S. Malamis, K.J. Haralambous, Industrial wastewater pre-treatment for heavy metal reduction by employing a sorbent-assisted ultrafiltration system, Chemosphere 82 (4) (2011 January) 557–564.

[53] S.M. Ritchie, D. Bhattacharyya, Membrane-based hybrid processes for high water recovery and selective inorganic pollutant separation, J. Hazard. Mater. 92 (1) (2002 May 3) 21–32.

[54] H. Kurama, J. Poetzschke, R. Haseneder, The application of membrane filtration for the removal of ammonium ions from potable water, Water Res. 36 (11) (2002 June) 2905–2909.

[55] I. Koyuncu, M. Yazgan, Application of nanofiltration and reverse osmosis membranes to the salty and polluted surface water, J. Environ. Sci. Health A Tox. Hazard. Subst. Environ. Eng. 36 (7) (2001) 1321–1333.

[56] S.W. Myung, I.H. Choi, S.H. Lee, I.C. Kim, K.H. Lee, Use of fouling resistant nanofiltration and reverse osmosis membranes for dyeing wastewater effluent treatment, Water Sci. Technol. 51 (6–7) (2005) 159–164.

[57] C.P. Leo, W.K. Chai, A.W. Mohammad, Y. Qi, A.F. Hoedley, S.P. Chai, Phosphorus removal using nanofiltration membranes, Water Sci. Technol. 64 (1) (2011) 199–205.

[58] R. Rautenbach, K. Vossenkaul, T. Linn, T. Katz, Waste water treatment by membrane processes – new development in ultrafiltration, nanofiltration and reverse osmosis, Desalination 108 (1–3) (February 1997) 247–253.

[59] X.-Y. Li, H.P. Chu, Membrane bioreactor for the drinking water treatment of polluted surface water supplies, Water Res. 37 (19) (November 2003) 4781–4791.

[60] S.M. Ruel, J.M. Choubert, M. Esperanza, C. Miège, P.N. Madrigal, H. Budzinski, K. Le Ménach, V. Lazarova, M. Coquery, On-site evaluation of the removal of 100 micro-pollutants through advanced wastewater treatment processes for reuse applications, Water Sci. Technol. 63 (11) (2011) 2486–2497.

[61] F.J. Simmons, D.H. Kuo, I. Xagoraraki, Removal of human enteric viruses by a full-scale membrane bioreactor during municipal wastewater processing, Water Res. 45 (9) (2011 April) 2739–2750.

[62] G. Laera, J.B. Lopez, Application of sequencing batch membrane bioreactors (SB-MBR) for the treatment of municipal wastewater, Water Sci. Technol. 64 (2) (2011) 391–396.

[63] S. Mondal, S. Ranil Wickramasinghe, Produced water treatment by nanofiltration and reverse osmosis membranes, J. Membr. Sci. 322 (1) (1 September 2008) 162–170.

[64] I. Senta, M. Matošić, H.K. Jakopović, S. Terzic, J. Curko, I. Mijatović, M. Ahel, Removal of antimicrobials using advanced wastewater treatment, J. Hazard. Mater. 192 (1) (2011 August 15) 319–328.

[65] A. Sperlich, D. Warschke, C. Wegmann, M. Ernst, M. Jekel, Treatment of membrane concentrates: phosphate removal and reduction of scaling potential, Water Sci. Technol. 61 (2) (2010) 301–306.

[66] G. Owen, M. Bandia, J.A. Howella, S.J. Churchouseb, Economic assessment of membrane processes for water and waste water treatment, J. Membr. Sci. 102 (15 June 1995) 77–91.

[67] Y. Kaiya, Y. Itoh, K. Fujita, S. Takizawa, Study on fouling materials in the membrane treatment process for potable water, Desalination 106 (1996) 71–77.

[68] J.H. Kweon, D.F. Lawler, Fouling mechanisms in the integrated system with softening and ultrafiltration, Water Res. 38 (2004) 4164–4172.

[69] S.R. Gray, C.B. Ritchie, B.A. Bolto, Effect of fractionated NOM on low pressure membrane flux declines, Water Sci. Technol. 4 (2004) 189–196.

[70] A. Maartens, P. Swart, E.P. Jacobs, Feed water pretreatment methods to reduce membrane fouling by natural organic matter, J. Membr. Sci. 162 (1999) 51–62.

[71] N. Her, G. Amy, C. Jarusutthirak, Seasonal variation of nanofiltration (NF) foulants: identification and control, Desalination 132 (2000) 143–160.

[72] T. Carroll, S. King, S.R. Gray, B.A. Bolto, N.A. Booker, The fouling of microfiltration by NOM after coagulation treatment, Water Res. 34 (1999) 2861–2868.

[73] J. Cho, G. Amy, J. Pellegrino, Membrane filtration of natural organic matter: factors and mechanisms affecting rejection and flux decline with charged ultrafiltration (UF) membrane, J. Membr. Sci. 164 (2000) 89–110.

[74] F.H. Frimmel, F. Saravia, Gorenflo, NOM removal from different raw waters by membrane filtration, Water Sci. Technol. 4 (2004) 165–174.

[75] L. Mo, X. Huang, Fouling characteristics and cleaning strategies in a coagulation–microfiltration combination process for water purification, Desalination 159 (2003) 1–9.

[76] K.J. Howe, K.P. Ishida, M.M. Clark, Use of ATR/FTIR spectrometry to study fouling of microfiltration membranes by naturals waters, Desalination 147 (2002) 251–255.

[77] P. Park, C.H. Lee, S.J. Choi, K.H. Choo, S.H. Kim, C.H. Yoon, Effect of the removal of DOMs on the performance of a coagulation–UF membrane system for drinking water production, Desalination 145 (2002) 237–245.

[78] D. Violleau, H.E. Tome, H. Habarou, J.P. Croue, M. Pontie, Fouling studies of a polyamide nanofiltration membrane by selected natural organic matter: an analytical approach, Desalination 173 (2004) 223–238.

[79] S. Xia, J. Nan, R. Liu, G. Li, Study of drinking water treatment by ultrafiltration of surface water and its application to China, Desalination 170 (2004) 41–47.

[80] C. Lin, T. Lin, O.J. Hao, Effects of humic substance characteristics on UF performance, Water Res. 34 (4) (1999) 1097–1106.

[81] T. Leiknes, H. Odegaard, H. Myklebust, Removal of natural organic matter (NOM) in drinking water treatment by coagulation–microfiltration using metal membranes, J. Membr. Sci. 242 (2004) 47–55.

[82] S. Mozia, M. Tomaszewska, Treatment of surface water using hybrid processes – adsorption on PAC and ultrafiltration, Desalination 162 (2004) 23–31.

[83] L. Fan, J.L. Harris, F.A. Roddick, N.A. Booker, Influence of the characteristics of natural organic matter on the fouling of microfiltration membranes, Water Res. 35 (18) (2001) 4455–4463.

[84] H. Lee, G. Amy, J. Cho, Y. Yoon, S.H. Moon, S.I. Kim, Cleaning strategies for flux recovery of an ultrafiltration membrane fouled by natural organic matter, Water Res. 35 (2001) 3301–3308.

[85] W. Yuan, A.L. Zydney, Humic acid fouling during microfiltration, J. Membr. Sci. 157 (1999).

[86] G.F. Crozes, J.G. Jacangelo, C. Anselme, J.M. Laine, Impact of ultrafiltration operating conditions on membrane irreversible fouling, J. Membr. Sci. 124 (1997) 63–76.

[87] N. Lee, G. Amy, J.P. Croue, H. Buisson, Identification and understanding of fouling in low pressure membrane (MF/UF) filtration by natural organic matter (NOM), Water Res. 38 (2004) 4511–4523.

[88] M.L. Manttari, J. Puro, N. Jokinen, M. Nystrom, Fouling effect of polysaccharides and humic acid in nanofiltration, J. Membr. Sci. 165 (2000) 1–17.

[89] M.E. Walsh, G.A. Gagnon, Evaluating membrane processes for drinking water treatment design. AEESP, Case Studies Compilation 2006, pp. 70–82.

6 Water Treatment by Electrical Technologies

CHAPTER OUTLINE

6.1 Introduction

The development of new technologies to treat wastewater has become a principal issue for a sustainable use of water resources as population of the world as well as the pollution

Environmental Water, http://dx.doi.org/10.1016/B978-0-444-59399-3.00006-4

increases rapidly. Strategies of water protection generally include development of new or improved wastewater processes, which have no or minor effects on the nature, i.e. the developed wastewater methods should be eco-friendly. In these situations, electrochemical processes used for treatment of wastewater play an important role. Electricity was first used to treat wastewater in 1889 [1] but due to expensiveness and shortage of electricity at that time it did not find wide application of these processes. In the current world order with the ever-increasing water standards and strict regulation of wastewater discharge and availability of electricity from hydrothermal to nuclear sources has increased the demand for the electrochemical processes. Many researchers are attempting to use electrochemical methods for the treatment of wastewater [2–5]. In the current world scenario, there is a great demand for these electrochemical methods for the treatment of wastewater as these methods provide energy efficiency, flexibility, amenability to automation, compatibility, and cost-effectiveness. A number of review articles are available regarding the application of electrochemical methods for environmental pollutants [6–8]. Moreover, electrochemical methods have been successfully applied in the purification of several industrial wastewaters [3,9–11] landfill leachate [12], and domestic sewage [13]. Recently, electrochemical techniques are gaining significance for the treatment of wastewater containing dyes [14,15], arsenic [16,17], chromium [18], phenols [19], etc. Electrochemical methods have been used as chemical coagulation processes to remove color and cloudiness from colored industrial wastewater. In this regard, it has been observed that the electrochemical process generated numerous flocs, which permitted to attain high efficiency in clearing wastewater [20]. The electrochemical methods have attracted a great deal of interest because of their flexibility and environmental compatibility, and which have made possible the treatments of liquids, gases, and solids. In fact, the main reagent of all the electrochemical process is the electron, which is considered as a clean reagent [6]. The electrochemical treatment of wastewater containing dye has drawn a considerable amount of attention due to its versatility and environmental compatibility through electron reactions [21].

The performance of any electrochemical process depends upon various parameters that influence efficacy of that method. The important parameters [22] determining the electrochemical performance will be as follows:

A. **Electrode potential and current density:** Controls the reaction/reaction rate at the electrodes and thus controls the efficiency of the method.
B. **Current distribution:** Should be as homogeneous as determines the spatial distribution of the consumption of reactants.
C. **Mass transport:** A high mass transport leads to uniform distribution of the pollutants around the electrodes and thus helps in increasing the efficiency of the process.
D. **Electrochemical cell:** The design, dimension, and the distance between the electrodes affect the efficiency of the electrochemical process.
E. **Electrolysis medium:** The type of electrolyte and its concentration, pH, and temperature of the medium affect the efficiency of the process.

F. Electrode materials: The choice of electrodes depends upon the type of pollutant; generally, electrode should be cheap, stable in the electrolysis medium, and exhibit high activity toward organic oxidation and low activity toward secondary reactions (e.g. oxygen evolution).

There are three main electrochemical processes for the water/wastewater treatment which are discussed below.

6.2 Electroflotation

EF is a simple process that floats pollutants to the surface of a water body by the impact of bubbles of hydrogen and oxygen gases generated from water electrolysis [23]. In other words, flotation of pollutants or any other substance to the surface with the adhesion/adsorption on the gas molecules produced during the electrolysis.

At anode:

$$2H_2O \rightarrow O_2 + 4H^+ + 4e^- \tag{6.1}$$

At cathode:

$$2H^+ + 2e^- \rightarrow H_2 \tag{6.2}$$

Overall reaction:

$$2H_2O \rightarrow O_2 + 2H_2 \tag{6.3}$$

Therefore, the reactions at the anode and cathode are oxygen evolution and hydrogen evolution reactions, respectively. Moreover, Eqn (6.3) also dictates that the amount of twofold hydrogen gas generation as compare to oxygen gas generation. The rate of gas generation can be calculated by applying Faraday's law:

$$Q = \frac{IV_O}{n\mathrm{F}}$$

where Q is the gas (hydrogen or oxygen) generating rate (L/s) at the normal state; V_O is the molar volume of gases at the normal state (22.4 L/mol), F the Faraday's constant (96,500 C/mol. electrons), n = number of the electrons transfer number of hydrogen and oxygen (2 mol electrons per mole of H_2 and 4 mol electrons per mole of O_2).

Thus, the overall generation of gases can be calculated:

$$Q_g = Q_H + Q_O$$

$$Q_g = \frac{IV_O}{n_H\mathrm{F}} + \frac{IV_O}{n_O\mathrm{F}}$$

By inserting the values of V_O, n_H, n_O, and Faraday's constant F, we get

$$Q_g = 1.74 \times 10^{-4} I$$

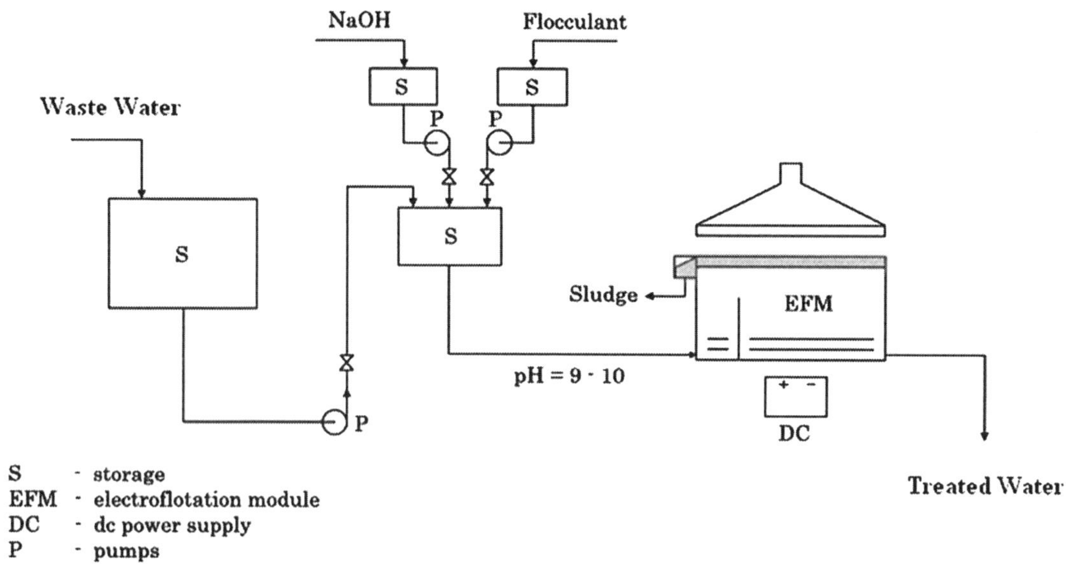

FIGURE 6-1 Schematic diagram of electro-flocculation.

Where Q_g is the total of all gases generation rate and I is the electrolysis current. A schematic diagram of electro-flocculation is shown in Figure 6-1.

6.2.1 Wastewaters Treated by EF

It was first reported by Elmore in 1904 for EF of minerals from ores [24] and is still one of its primary usages in minerals recovery [25–28]. Moreover, EF is very effective in water and wastewater treatment for the separation of metal ions [29–38]. EF is found effective in treating oil-wastewater, oil-water effluent [39–45], groundwater decontamination [46], apple juice and food processing wastewater [47–49], sewage water and pit waters [50–52], colloidal and suspended particles [53–55], laundry and textile wastewater [56–58], wastewater of paper industry [59,60], and many other water and wastewaters like water cooling lubricant [61], fats containing solutions [62], dairy wastewater [63], etc. Nowadays, EF method hyphenated with EC is used for the treatment of various wastewater like defluoridation of drinking water [64–66], removal of chromium [38], dye [58], and laundry wastewater[67] and industrial wastewater [68].

6.2.2 Factors Affecting Electroflotation

The performance of any electrochemical method system depends upon removal efficiency of the contaminant and the power consumptions. The removal efficiency of an EF system depends on the size of the gas bubbles formed during the experiment. While the power consumption is directly dependant on the operating conditions such as current density, water conductivity, etc. and more importantly the cell design and electrode materials. Important factors that influence the EF systems are summarized below:

6.2.2.1 EF Cell Designs

The main parts of any EF unit consist of two electrodes and a power supply. The configuration of the electrodes may be horizontal or vertical and are usually placed at the bottom of the cell. The horizontal placement is the most popular choice. There are generally two types of cell design, i.e. single or two stage.

6.2.2.2 Effect of pH

The gas bubbles size depends on the wastewater pH, current density, and the electrode material. Two individual groups Brandon and Kelsall [69] and Glembotskii et al. [70] reported that during EF the size of the oxygen bubble increases with an increase in the pH, while the bubbles of hydrogen gas was smallest at neutral pH. Moreover, the size of hydrogen gas molecules was bigger in size in acidic range than in the basic pH range. Both anodic and cathodic material affects the size of the hydrogen molecules. Llerena et al. [71] studied the effect of pH on the size of the gas bubbles produced during the recovery of the sphalerite by EF. The best recovery occurs in the pH range of 3–4, where the size of the hydrogen gas bubbles was smallest. Decrease or increase in the pH range increases the size of the hydrogen bubble when all the other experimental parameters were identical and thus thereby decreases the recovery of the sphalerite. Other workers [72,73] also reported that the recovery of particles in the diameter range of 1–10 μm is increased with decreasing bubble size of the gas generated during the EF process, which is mainly due to the better collision efficiency between the pollutant particles and the bubbles.

6.2.2.3 Current Density

There are different opinions regarding the influence of current density on the size of the bubble. Some workers [74–77] have reported the trend with increase in bubble size with increase in the current density. They accredited this to bubble coalescence at higher current densities. Contrarily, other workers reported the inverse of this trend [78–80]. Moreover, it has also been reported that mean bubble size of the gas increases with smoothed surface of the electrode while the rough surface shows decrease of the bubble when the current density on the electrodes increases [81]. Apart from size of bubble, the bubble flux, i.e. the number of bubbles presented per second per unit cross section area of the EF cell, also plays a role in the recovery of mineral flotation and different sized particles [82].

6.2.2.4 Arrangement of the Electrodes

The EF electrodes can be arranged in different patterns. Generally, a plate electrode is installed at the bottom, while a screen electrode is fixed at 10–50 mm above the plate electrode as shown in Figure 6-1 [36,71,83,84]. In this type of electrode system, only the

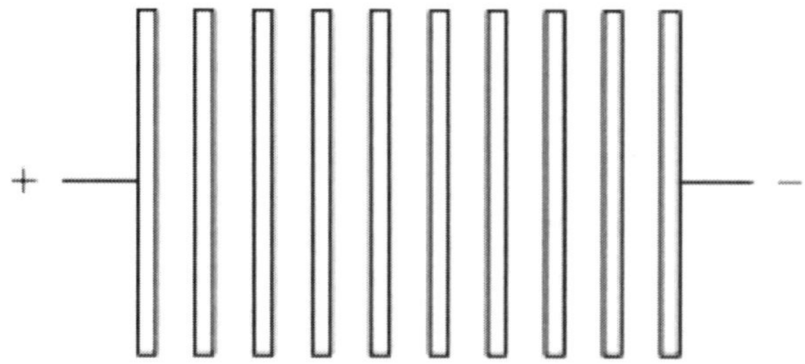

FIGURE 6-2 Arrangement of electrodes in vertical EF.

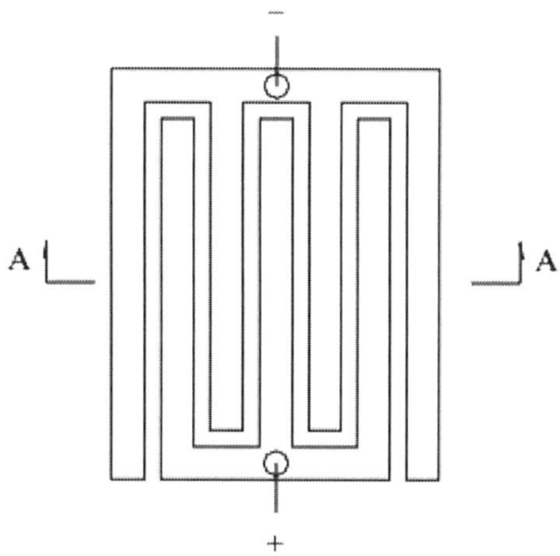

FIGURE 6-3 Arrangement of electrodes in fork like EF.

upper screen electrode directly interacts with the water flow, while the bottom screen electrode does not interact with the flow directly. The bubbles generated at the bottom electrode cannot be dispersed immediately into the wastewater being treated. Thereby decreases the availability of the effective small bubbles, but also increases the possibility of breaking the flocs, affecting the flotation efficiency. Furthermore, if the conductivity of wastewater is low, energy consumption will be unacceptably high due to the large inter-electrode spacing required for preventing the short circuit between the upper flexible screen cathode and the bottom anode. In such types of electrodes system, the maintenance of the electrodes system is also a problem as screen electrode is easy to be twined by undesirable deposits such as fabric and thus difficult to clean once it twined. The arrangement of the electrodes can also be arranged vertically as shown in the Figures 6-2 and 6-3 [47,85,86].

For the arrangement of electrodes as shown in the Figure 6-3, they are easy to fix as both the anode and cathode are of plate shape. While in the other case of electrode, a fork like arrangements of the anode and cathode (Figure 6-3) are placed on the same plane. The efficiency of the EF system with these forks like arrangement increases due to the

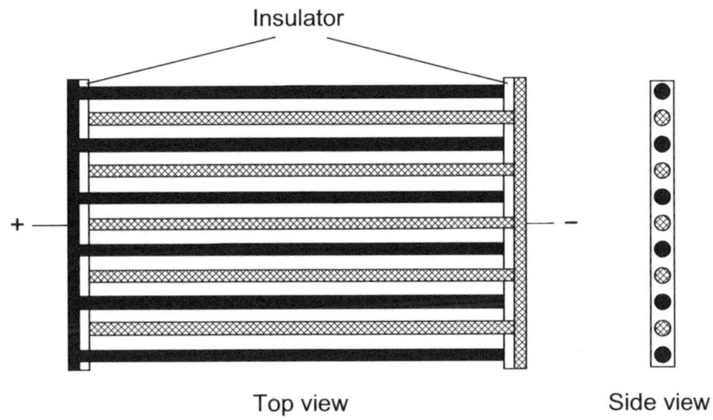

FIGURE 6-4 Open electrodes configuration in EF.

quick dispersion of the generated bubbles into the water flow and attached with the pollutants. In these types, there is a direct contact of the water flow with the anode as well as the cathode. As the quick dispersion of the gas bubbles occurs therefore the size of the electrodes remains small and these smaller bubbles provide larger surface area for particle attachment; moreover, the bubbles generated at both the electrodes also enhance the efficiency of the system.

Another open electrodes configuration is shown in the Figure 6-4 [38] that is mostly used in the flotation of oil and other suspended solid [87]. In this type, the distance between the electrodes is quite small (up to 2 mm), which helps in the energy conservation of the EF system as the inter-electrode gap is directly proportional to the ohmic potential drop of the EF cell thus reducing this distance is of great importance for reducing the electrolysis energy consumption.

6.2.2.5 Electrodes Materials

The arrangement and the type of electrode material are the back bone of the EF unit. The most common types of electrode material used for the EF are iron, aluminum, stainless steel, graphite, and lead oxide [36,41,71,88,89,90,91] which are inexpensive and easily available. Aluminum and stainless steel electrodes are mostly used in EC-EF hyphenated systems as they are anodically soluble but are not the choice for the EF systems as they produce the bubbles which are larger in size dissolving coarse surface of the electrodes. For the insoluble purposes of the EF method, graphite and lead oxide electrodes are used. Although both these electrodes are inexpensive and show high oxygen evolution during EF process, they have less durability. Moreover, the lead oxide electrode also creates the risk to generate toxic Pb^{2+} ions in the treated wastewater. Some other scientists use platinum [46,82] electrodes as they are more stable than the graphite and lead oxide, but they are highly expensive and thus are not used at large scale. Nowadays DSA (dimensionally stable anode) are used in the EF technique. These DSA use conductive metal oxides like RuO_2, IrO_2, as electrocatalyst and non conductive metal oxides like TiO_2, Ta_2O_5, ZrO_2, Nb_2O_5, as dispersing or stabilizing agents, which are coated on valve metal substrates (Ti, Ta, Zr, W, Nb, Bi) with a thermal decomposition method [92,93]. TiO_2–RuO_2 DSA has a tendency to produce a large amount of chlorine but they are less durable for the evolution of oxygen [94] due to which IrOx-based DSA have gain popularity as they are about 20 times more durable than the RuO_2 [95]. Sometimes high electro thermal electrode is required for the evolution of the oxygen, for this purpose Ti/IrO_x–Sb_2O_5–SnO_2 DSA were used [96,97]. These contain about 10 mol% of IrO_x in the oxide coating and have 9 years of working life at the high current density of 1000 A/m^2 in the high acidic medium. A 2.5 mol% of IrO_x in the oxide coating gives the same results and having the same life if used at low current density [98].

6.3 Electrocoagulation

EC is a simple technique where the flocculating agent is generated by oxidation of a sacrificial anode, which is generally made up of iron or aluminum metal. In this process, treatment is done without adding any chemical coagulant or we can say it is generated by oxidation of the anode material when the electricity is passed through the EC cell [99]. The chemical reaction takes place for the iron and aluminum electrodes are given below:

For aluminum anode:

$$Al \rightarrow Al^{3+}_{(aq)} + 3e^- \quad (6.1)$$

Under alkaline conditions

$$Al^{3+}_{(aq)} + 3OH^-_{(aq)} \rightarrow Al(OH)_{3(s)} \quad (6.2)$$

Under acidic conditions

$$Al^{3+}_{(aq)} + 3H_2O_{(l)} \rightarrow Al(OH)_{3(s)} + 3H^+_{(aq)} \quad (6.3)$$

For iron anode:

$$Fe \rightarrow Fe^{2+} + 2e^- \quad (6.4)$$

Under alkaline conditions

$$Fe^{2+}_{(aq)} + 2OH^-_{(aq)} \rightarrow Fe(OH)_{2(s)} \quad (6.5)$$

Under acidic conditions

$$4Fe^{2+}_{(aq)} + O_2 + 2H_2O_{(l)} \rightarrow 4Fe^{3+} + 4OH^-_{(aq)} \quad (6.6)$$

Reaction at the cathode:

$$2H_2O_{(l)} + 2e^- \rightarrow H_{2(g)} + 2OH^-_{(aq)} \quad (6.7)$$

$$8H^+_{(aq)} + 8e^- \rightarrow 4H_{2(g)} \quad (6.8)$$

The whole EC process can be dived into three stages, first the formation of coagulants at the sacrificial anode when the electricity is passed through the EC cell, the generated coagulant is nothing but the nascent metal ions like Al^{3+} or Fe^{2+} or the hydroxides of these metal ions. Secondly, the destabilization of the contaminant particles, the breaking of emulsions, etc. Finally, the accumulation of the destabilized contaminants particle to form flocs. Moreover, it was also seen that water is also electrolyzed in a parallel reactions during the process thus producing small bubbles of oxygen at the anode and hydrogen at the cathode. These bubbles attract the flocculated particles and float the flocculated pollutants to the surface. Therefore, EC is also used in hyphenation with EF [38,58,64–68]. In addition, the following various other reactions may also take place in the EC cell-like cathodic reduction of impurities present in water, coagulation of colloidal particles, electrophoretic migration of the ions in solution, EF of the coagulated particles by O_2 and H_2 bubbles produced at the electrodes, reduction of metal ions at the cathode, and other electrochemical and chemical processes [100].

6.3.1 Wastewater Treated by EC

EC was found useful in the treatment of various wastewaters. It has been employed in treating wastewaters from textile [101–104], chemical fiber wastewater [105], oily wastewater [106], clay suspension [107], petroleum, tar sand and oil wastewater [108], restaurant wastewater [89,109], carpet wastewater [110], sewage water [109], oil–water emulsion [111,112], nitrite [113], and dye stuff [113,114]. The EC was also used for the removal of solution containing various metal ions like, Cu, Cr, Ni, In, Co, Mn, Zn, Ag [115–119]. It can be used to remove irons, silicates, humus, dissolved oxygen, etc [120]. Recently, many scientific groups around the world use EC method for the removal of arsenic [121–125], as it is carcinogenic in nature and about 150 million people are at risk around the world [126]. It is also used for coagulating the colloidal found in natural water so that reduces the turbidity and color. It is also used in the removal or destruction of algae or microorganisms [127].

6.3.2 Factors Affecting EC

6.3.2.1 Electrode Materials

The electrode materials employed in the EC method are usually made up of aluminum or iron metal, other metal electrodes like titanium, copper, zinc, etc. are also used. The efficiency of the EC cell decreases when the sludge or the contaminants particles are deposited on the anode. Therefore, it is necessary to rinse the surface of electrodes with dilute acid, as there is a definite amount of metal ions required to remove a given amount of pollutants. The electrode systems used in the EC may be monopolar or bipolar; moreover to increase the efficiency, sometime periodic polarity reversal of the electrodes is also done [128].

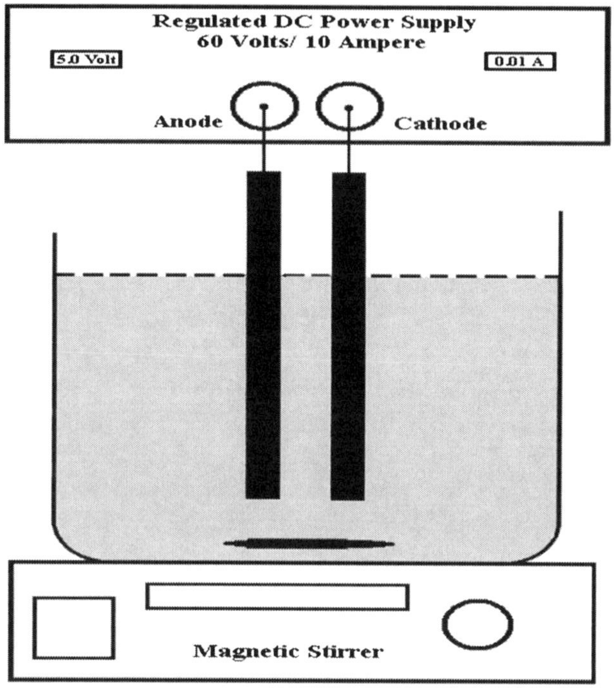

FIGURE 6-5 Setup of EC process.

6.3.2.2 EC Cell Design

A schematic diagram of EC is given in Figure 6-5. It consists a chamber and two electrodes. Both the electrodes are remaining attached with the power supply.

6.3.2.3 Current Density

The amount of the supply of current to the electrodes determines the amount of the oxidation of the metal ion generated at the anode. Current density of the EC system is the only parameter that is controlled directly. It also controls both coagulant dosage and bubble generation rates and strongly influences both solution mixing and mass transfer at the electrodes. Thus, the current density is the one of the main operational factor of the EC method. The highest allowable current density may not be the most efficient mode as it results in heating up the wastewater and therefore wasting an amount of electricity. Chen et al. [129] suggested that for EC system to operate for a long duration of time without maintenance, its current density should be in the range of 20–25 Am^{-2}. The current density selection should be made with other operating parameters such as pH, temperature, as well as flow rate to ensure a high current efficiency (CE).

6.3.2.4 Effect of Water Conductivity

The conductivity of the water directly influences the CE. It is seen when the conductivity of the water decreases, the high applied potential is needed for the passivation of electrode and thus increases the treatment cost of wastewater. Generally, NaCl was added to increase the electrolytic conductivity of the wastewater, the addition of table salt also helps in the water disinfection [130]. It is also reported that it shows a negative effect on carbonate and sulfate which lead to the formation of oxide layer, the formation of oxide layer decreases the efficiency. The study of KNO_3, $NaNO_3$, KCl, and NaCl on the removal of arsenic was studied by Pinisakul et al. [131] and reported that the efficiency of arsenic removal was lower in case of nitrate salts as compared to chloride salts. The reason for this effect was the less production of total dissolved solids (TDS) in case of nitrate salts than the chloride salts. In another study on arsenic removal by EC, Lakshmipathiraj et al. [16] studied the effects of NaCl, Na_2SO_4, and $NaNO_3$. It is found that in case of NaCl electrolyte, the removal was much higher 98% as compared to 80% and for Na_2SO_4 and $NaNO_3$. Both the salts Na_2SO_4 and $NaNO_3$ help in the production of passive film around the electrodes, thereby decreasing the efficiency of the EC method.

6.3.2.5 Effect of pH on EC

The pH of solution is one of the important role in electrochemical process. The treatment performance depends on the nature of the pollutants with the best pollutant removal found near pH of 7 but the power consumption is, however, higher at neutral pH due to the variation of conductivity. The formation of hydroxides after the dissolution of any metal ions depends upon the pH of the solution. For example during the dissolution of aluminum, the formation of various hydroxides depends upon the pH of the solution. In the pH range of 4–9, the formation of $Al(OH)^{2+}$, $Al(OH)_2^{+}$, $Al_2(OH)_2^{4+}$, $Al(OH)_3$, and $Al_{13}(OH)_{32}^{7+}$ takes place. The surface of these hydroxides has large amounts of positive charge, which helps in the adsorption and net catching reaction. At pH >10, $Al(OH)^{-}{}_4$ is dominant species and which shows less coagulation as compared to the former species. At low pH, Al^{3+} is the most dominant species and shows almost no coagulation effect [132]. It is also shown that when the conductivity of the solution is high, pH effect is not much significant in this situation. It is interesting to note that the pH of the solution slightly increases during the EC [133] due to the formation of H_2 at the cathode and thereby increases the OH^- concentration in the solution.

6.3.3 Advantages of EC

EC requires simple electrochemical cell and is easy to operate. The treated wastewater was clear, colorless, and odorless water. The sludge formed by EC is very much less as compared to the chemical coagulation. The flocs/waste formed during the process was readily settable as it is composed of mainly metallic oxides/hydroxides. Flocs formed by EC are similar to chemical floc, except that EC floc tends to be much larger, contains less bound water, is acid resistant, and more stable and, therefore, can be separated faster by filtration. Moreover, the EC produces wastewater has less TDS content as compared with chemical treatments. Or we can say EC process has the advantage of removing the smallest colloidal particles, because the applied electric field sets them in faster motion, thereby facilitating the coagulation. The gas bubbles produced during EC can carry the pollutant to the top of the solution where it can be more easily concentrated, collected, and removed. The most important advantage of this process, it avoids uses of chemicals, and so there is no problem of neutralizing excess chemicals and no possibility of secondary pollution caused by chemical substances added at high concentration as when chemical coagulation of wastewater is used.

6.4 Electrooxidation

The results of electrochemical degradation of some organic and inorganic compounds in effluent streams started a new filed in electrochemistry, known as Electrooxidation [134]. The main advantage of this technology is their efficiency to remove pollutants from wastewater without adding any chemical and at lower cost than any technology used, which is an important issue in water treatment. After 20 years of rigorous works and development led by different scientists, we are still looking into the following factors to make this technology better:

(a) Factors affecting electrooxidation process,
(b) Exact mechanism and kinetics underlying this process,
(b) Enhancement of the catalytic and chemical stabilities of electrodes in varying environment,
(c) Efficiency of different electrodes for varying organic pollutants.

However, although much work have been done on the anodic materials, less results have been reported for cathodic materials. The role of cathodic materials can be understood by the fact that by changing the cathodic part by different materials, the destruction of 4-chloro phenol also get influenced [135], which clearly indicates the role of both electrodes in this process. The flow sheet of electrooxidation is depicted in Figure 6-6.

6.4.1 Direct Anodic Oxidation

As per the name implies, the pollutants in water sample are oxidized by using current at the anodes. The oxidation is done by generating either physically adsorbed "active oxygen" (adsorbed hydroxyl radicals, $^{\bullet}OH$) or chemisorbed "active oxygen" (oxygen in the oxide lattice, MO_{x+1}) [2]. The complete combustion of pollutants occurs when electro-oxidation occurs through physically adsorbed "active oxygen", while chemisorbed "active oxygen" generates various products:

$$R + MO_x(^{\bullet}OH)_z = CO_2 + zH^+ + ze^- + MO_x$$

$$R + MO_{x+1} = RO + MO_x$$

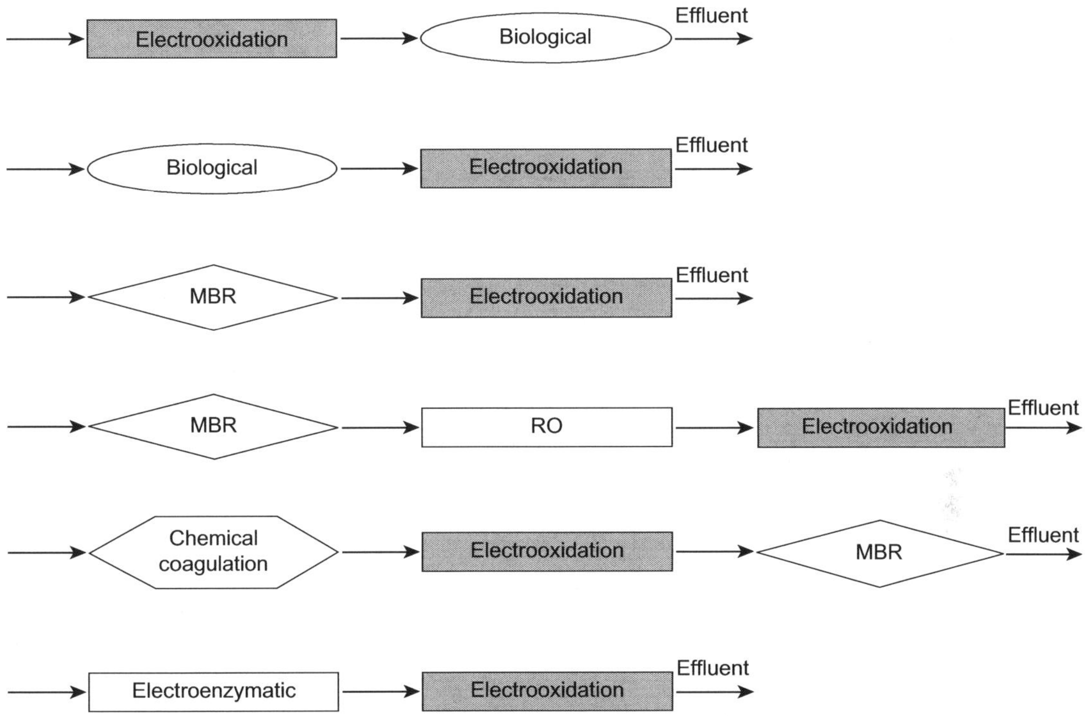

FIGURE 6-6 Flow sheet of electrooxidation.

Because oxygen evolution can also take place at the anode, high overpotential for O_2 evolution are required in order for above said reactions to proceed with high CE. Otherwise, most of the current supplied will be wasted to split water. The anodic oxidation does not need to add a large amount of chemicals to wastewater or to feed O_2 to cathodes, with no tendency of producing secondary pollution and fewer accessories required. These advantages make anodic oxidation more attractive than other electro-oxidation processes. The important part of an anodic oxidation process is obviously the anode material. Anode materials investigated include glassy carbon [136], Ti/RuO_2, Ti/Pt–Ir [137], fiber carbon [138], MnO_2 [139,140], Pt–carbon black [141,142], porous carbon felt [143], stainless steel, and reticulated vitreous carbon [144,145]. The anodes that were studied extensively are graphite, Pt, PbO_2, IrO_2, TiO_2, SnO_2, and diamond film. However, none of them have sufficient activity and at the same time stability. We will discuss them in more details subsequently.

6.4.1.1 Overpotential of Oxygen Evolution

The anodic activity depends on the value of overpotential of oxygen evolution. It has been found that IrO_2, Pt, and graphite show much smaller values of overpotential of oxygen evolution. It means that effective oxidation of pollutants on these anodes occurs only at very low current densities or in the presence of high concentrations of chlorides or metallic mediators. When the current density is high, significant decrease of the CE is expected from the production of oxygen. The boron-doped diamond (BDD) film on titanium substrate [146] or other valve metals as in DiaChem electrodes [147] gives the highest value of oxygen evolution overpotential. Thus, anodic oxidation can take place on its surface at significantly high current density with minimal amount of oxygen evolution side reaction. This leads to an effective and efficient process. It is indeed the most active anode for oxidation of various pollutants as discussed in the following sections.

6.4.1.2 Performance of Anodic Oxidation

The important parameters to check the performance of the anodic oxidation are current density and the CE. An important parameter proposed by the Comninellis and Plattner [148] was electrochemical oxidability index (EOI) to differentiate the performances of various electrodes. EOI can be defined as the mean CE from the initial concentration of pollutant to the time is nearly zero, τ. EOI can be calculated as

$$\mathrm{EOI} = \frac{\int_0^{\tau} \mathrm{ICE}\,\mathrm{d}t}{\tau}$$

where ICE is the instantaneous CE (CE at the time of electrooxidation). EOI calculated for any electrode for are always low and lies within the range of 0.05–0.58 for electrochemically degrading various benzene derivatives on Pt anode [148]. Graphite anode shows a high CE (70%) at low current densities which ranging from 0.03 to 0.32 A/m^2, but when the current densities increased in the range of 10–100 A/m^2, the CE values decreased to 6–17% [149,150]. PbO_2 is the one of the most used anodic material in electrooxidation. The activity of the PbO_2 is increased by doping with by Bi, Fe, and Ag [151–153]. PbO_2 is used for the oxidation of aniline having a good value of CE and current density which lies in the range of 80–160 A/m^2. Moreover, this anodic material is inexpensive and very effective in oxidizing various pollutants [154,155]. But the only disadvantage of this material is the formation of Pb^{2+} from the electrochemical oxidation. Pure SnO_2 is an n-type semiconductor with a band gap of about 3.5 eV. SnO_2 and other similar oxides show a very high resistivity at room temperature and therefore cannot be used as an electrode material directly. However, conductivity can be improved significantly by doping Ar, B, Bi, F, P, and Sb [156–162]. Sb is the most common dopant of SnO_2 oxides and used as transparent electrodes in high-efficiency solar cells, gas detectors, far IR detectors, and transparent heating elements. Comninellis [163] measured the CE of SnO_2–Sb_2O_5 to be 0.58 for 71% degradation of phenol [164], while the values for PbO_2, IrO_2, RuO_2, and Pt are, respectively, 0.18, 0.17, 0.14, and 0.13 at same experimental conditions. Grimm et al. [165] studied the oxidation of phenol on SnO_2–Sb_2O_5 and PbO_2 showed that and reported that SnO_2–Sb_2O_5 was more active than PbO_2. But Cossu et al. [166] reported that there is no much difference in the activity between the two materials in treating landfill leachate, which may results due the high concentration of chlorides in leachate waste. The stability of Sb_2O_5–SnO_2 electrodes was only 12 h under an accelerated life test performed at a current density of 1000 A/m^2 in 1 M H_2SO_4 solution, which reduces to few seconds under 10,000 A/m^2 and 3 M H_2SO_4 [167]. By adding IrO_2 into the SnO_2–Sb_2O_5 mixture to increase the life of it [96,97]. TiO_2 electrodes are stable at low current densities (below 30 A/m^2), but their lifetimes are significantly shortened when operated at high current densities [168].

6.4.2 Indirect Electrooxidation Processes

As we have already seen in the application of direct oxidation method, there are some other indirect electrooxidation method for the electrooxidation of wastewater. For an example, destruction of pollutants by the use of chlorine and hypochlorite generated anodically is well known method. The main advantage of this technology is its application for both organic and inorganic pollutants containing high level of chlorides [11,113,169–171]. The efficiency of this method mainly depends on the concentration of chloride in wastewater. Interestingly, where the excess of chlorine leads to the formation of chlorinated by-products, which interferes during the breakdown process and thus decreases the efficiency, but when low chloride concentration is present in the sample, it become necessary to add chloride salt to increase their efficiency [12,169,172,173]. Another method for the degradation of pollutant is the use of H_2O_2, which is generated in the cell electrochemically. In this type of cell, the anode part is made of Pb/PbO_2, Ti/Pt/PbO_2, or Pt and the cathodic part is made of carbon polytetrafluorethylene (PTFE) having O_2 feeding. Electro-Fenton reaction has been observed due to the presence of ferrous salt formed in situ, when iron is used as anodic material. While in other case, ferrous salt is added externally in the waste water [174–177].

UV irradiation does not lag too behind that above discussed process. Around 95 % of aniline was removed successfully by radiating the wastewater sample. The role of O_2 and O_3 is also interesting. It has been observed that by simply passing O_2 through the wastewater sample, it removes aniline in the presence of electric current, whereas O_3 generated during the process is also reported for the treatment of wastewater purpose [175,178]. As all the processes discussed can be used for both organic and inorganic materials, there are some materials which need special attention. Whenever there is mixed or hazardous material present in the wastewater, care should be taken. For this purpose, Farmer et al. [179] used an another approach, known as mediated electrooxidation. In this process, mediators, which are mainly metal ions like Ag^{2+}, Co^{3+}, Fe^{3+}, Ce^{4+}, and Ni^{2+}, are oxidized on an anode to a reactive, high valence state, which in turn attack organic pollutants directly and may also produce hydroxyl free radicals in the wastewater system that promote destruction of the organic pollutants. After that, they again come back to lower oxidation state at anode, and hence a cycle reaction starts which continues till the complete removal of pollutants [180–182]. This method is limited by the fact that its main component i.e. mediators are heavy metal ion which again pollutes water.

Conclusion

Electrochemical technologies have been investigated as the effluent treatment processes for many decades and with the advancement of science their importance have also increased. Nowadays, electrochemical methods are used in many fields from metal recovery to the wastewater treatment. Moreover, these methods can be applied at industrial scale. EF is mostly used in the mining process and it is also used in various wastewater treatment. The generated gas bubbles at the electrodes give much better results than either dissolved air flotation, sedimentation or even impeller flotation. EC has been used in the treatment of oil and grease, organic or inorganic pollutants, and the removal of various metal ions from its solutions. Electrooxidation process is mostly used for the treatment of organic pollutants. All these electrochemical processes were simple, compact, and easy to operate and inexpensive in nature.

References

[1] P.P. Strokach, Electrochem. Ind. Process Biol. 55 (1975) 375.

[2] C. Comninellis, Electrocatalysis in the electrochemical conversion/combustion of organic pollutants for wastewater treatment, Electrochim. Acta. 39 (11/12) (1994) 1857–1865.

[3] C. Comninellis, C. Pugarin, Anodic oxidation of phenol for waste water treatment, J. Appl. Electrochem. 12 (1991) 703–708.

[4] C. Comninellis, A. Nerini, Anodic oxidation of phenol in the presence of NaCl for wastewater treatment, J. Appl. Electrochem. 25 (1995) 23–28.

[5] K. Kusakabe, H. Nishida, S. Marooka, Y. Kato, Simultaneous electrochemical removal of copper and chemical oxygen demand using a packed-bed electrode cell, J. Appl. Electrochem. 16 (1986) 121–126.

[6] K. Rajeshwar, J.G. Ibanez, G.M. Swain, Electrochemistry and the environment, J. Appl. Electrochem. 24 (1994) 1077.

[7] A. Savall, Electrochemical treatment of industrial organic effluents, Chimia 49 (1995) 23.

[8] J.O.M. Bockris, R.C. Bhardwaj, C.L.K. Tennakoon, Electrochemistry of waste removal; a review, Analyst 119 (1994) 781–789.

[9] S. Liakou, S. Pavlou, G. Lyberatos, Ozonation of azo dyes, Water Sci. Technol. 35 (4) (1997) 279–286.

[10] H. Sharifian, D. Kirk, Electrochemical oxidation of phenol, J. Electrochem. Soc. 15 (1985) 921–924.

[11] J. Naumczyk, L. Szpyrkowicz, D.M. De Faverri, F. Zilio-Grandi, Electrochemical treatment of tannery wastewater containing high strength pollutants, Trans. IChemE 74 (B) (1996) 59–68.

[12] L.C. Chang, T.C. Wen, Indirect oxidation effect in electro-chemical oxidation treatment of land®ll leachate, Water Res. 29 (1995) 671–678.

[13] M. Della Monica, A. Agostizno, A. Ceglie, An electrochemical sewage treatment process, J. Appl. Electrochem. 10 (1980) 527–533.

[14] G. Bhaskar Raju, M. Thalamadai Karuppiah, S.S. Latha, S. Parvathy, S. Prabhakar, Treatment of wastewater from synthetic textile industry by electrocoagulation-electro-oxidation, Chem. Eng. J. 144 (2008) 51–58.

[15] M. Muthukumar, M. Thalamadai Karuppiah, G. Bhaskar Raju, Electrochemical removal of CI acid orange10 from aqueous solutions, Sep. Purif. Technol. 55 (2007) 198–205.

[16] P. Lakshmipathiraj, S. Prabhakar, G. Bhaskar Raju, Studies on the electrochemical decontamination of wastewater containing arsenic, Sep. Purif. Technol. 73 (2010) 114–121.

[17] Imran Ali, Tabrez A. Khan, Mohd Asim, Removal of arsenate from ground water by electrocoagulation method, Environ. Sci. Pollut. Res. Int. 19 (2012).

[18] P. Lakshmipathiraj, G. Bhaskar Raju, M. Raviatul Basariya, S. Parvathy, S. Prabhakar, Removal of Cr (VI) by electrochemical reduction, Sep. Purif. Technol. 60 (2008) 96–102.

[19] M. Murugananthan, G. Bhaskar Raju, S. Prabhakar, Removal of tannins and polyhydroxy phenols by electro-chemical techniques, J. Chem. Technol. Biotechnol. 80 (2005) 1188–1197.

[20] A.B. Riveiro, E.P. Mateus, L.M. Ottosen, G. Bech-Nielsen, Electrodialytic removal of Cu, Cr, and As from chromated copper arsenate-treated timber waste, Environ. Sci. Technol. 34 (2000) 781–788.

[21] N. Mohan, N. Balasubramanian, C. Ahmed Basha, Electrochemical oxidation of textile wastewater and its reuse, J. Hazard. Mater. 147 (2007) 644–651.

[22] D. Pletcher, F.C. Walsh, Industrial Electrochemistry, Chapman and Hall, London, 1982.

[23] G.B. Raju, Khangaonkar, Electroflotation—a critical review, Trans. Indian Inst. Met. 37 (1) (1984) 59–66.

[24] F.E. Elmore, A process for separating certain constituents of subdivided ores and like substances, and apparatus therefore, Br. Patent 13 (578) (1905).

[25] V.E. Nenno, V.I. Zelentsov, E.V. Mel'nichuk, A.M. Romanov, T.Y. Datsko, T.M. Radzilevich, Experience in operating a device for concentration of mineral raw material combining electroflotation and separation in a froth layer, Electronnaya Obrabotka Materialov. 6 (1988) 77–79.

[26] V.A. Glembotskii, A.A. Mamakov, E.S. Nenno, A.M. Romanov, V.E. Nenno, Study of Electroflotation of manganese muds. (ISSLEDOVANIE ELEKTROFLOTATSII MARGANTSEVYKH ILOV), Elektronnaya Obrabotka Materialov. 61 (1) (1975) 51–53.

[27] P. Hogan, A.T. Kuhn, J.F. Turner, Electroflotation studies based on cassiterite ores, Trans. Inst. Min. Metall. Sec. C: Miner. Process. Extr. Metall. 88 (1979) 83–87.

[28] R. Houot, Flotation of oxide minerals – recent progress. (FLOTTATION DES MINERAIS OXYDES - PROGRES RECENTS), Industrie Minerale, Les. Techniques(7) (1982) 361–367.

[29] V.A. Kolesnikov, V.I. Il'in, S.O. Varaksin, V.T. Shaturov, Electroflotation method and equipment for removing metals and organic contaminants from waste waters, Russ. J. Heavy Mach. 1 (1996) 37–38.

[30] B. Merzouk, B. Gourich, A. Sekki, K. Madani, M. Chibane, Removal turbidity and separation of heavy metals using electrocoagulation–electroflotation technique A case study, J. Hazard. Mater. 164 (2009) 215–222.

[31] V. Srinivasan, M. Subbaiyan, Electroflotation studies on Cu, Ni, Zn, and Cd with ammonium dodecyl dithiocarbamate, Sep. Sci. Technol. 24 (1–2) (1989) 145–150.

[32] G. Ramadorai, J.P. Hanten, Removal of molybdenum and heavy metals from effluents by flotation, Miner. Metall. Process. (1986) 149–154.

[33] V.I. Zelentsov, K.A. Kiselev, An investigation of separation of valuable components from solutions with electrical flotation, Elektronnaya Obrabotka Materialov. 4 (1986) 50–54.

[34] V.E. Nenno, V.I. Zelentsov, T.Ya. Datsko, E.E. Dvornikova, T.M. Radzilevich, Gold and silver sorption from cyanide solutions by activated coal and metal isolation by electroflotation, Electronnaya Obrabotka Materialov. 3 (1994) 42–44.

[35] S.M. Oussedik, A. Khelifa, Reduction of copper ions concentration in wastewaters of galvanoplastic industry by electroflotation, Desalination 139 (2001) 383.

[36] L. Alexandrova, T. Nedialkova, I. Nishkov, Electroflotation of metal ions in waste water, Int. J. Miner. Process. 41 (1994) 285–294.

[37] R.G. Casqueira, M.L. Torem, H.M. Kohler, The removal of zinc from liquid streams by electroflotation, Miner. Eng. 19 (2006) 1388–1392.

[38] P. Gao, X. Chen, F. Shen, G. Chen, Removal of chromium(VI) from wastewater by combined electrocoagulation–electroflotation without a filter, Sep. Purif. Technol. 43 (2005) 117–123.

[39] A.Y. Hosny, Separating oil from oil–water emulsions by electroflotation technique, Sep. Technol. 6 (1996) 9–17.

[40] M.Y. Ibrahim, S.R. Mostafa, M.F.M. Fahmy, A.I. Hafez, Utilization of electroflotation in remediation of oily wastewater, Sep. Sci. Technol. 36 (16) (2001) 3749–3762.

[41] C.C. Ho, C.Y. Chan, The application of lead dioxide-coated titanium anode in the electroflotation of palm oil mill effluent, Water Res. 20 (1986) 1523–1527.

[42] L.M. Balmer, A.W. Foulds, Separation oil from oil-in-water emulsions by electroflocculation/electroflotation, Filtr. Sep. 23 (11–12) (1986) 366–369.

[43] A.Y. Hosny, Separation of Oil from Oil/Water Emulsions using an Electroflotation Cell with Insoluble Electrodes, Filtr. Sep. (1992) 419–423.

[44] R.M. Bande, B. Prasad, I.M. Mishra, K.L. Wasewar, Oil field effluent water treatment for safe disposal by electroflotation, Chem. Eng. J. 137 (2008) 503–509.

[45] N.M. Mostefa, M. Tir, Coupling flocculation with electroflotation for waste oil/water emulsion treatment. Optimization of the operating conditions, Desalination 161 (2004) 115–121.

[46] C.P.C. Poon, Electroflotation for groundwater decontamination, J. Hazard. Mater. 55 (1997) 159–170.

[47] B.J. Hernlem, L.S. Tsai, Chlorine generation and disinfection by electroflotation, J. Food Sci. 65 (2000) 834–837.

[48] M. Araya-Farias, M. Mondor, F. Lamarche, S. Tajchakavit, J. Makhlouf, Clarification of apple juice by electroflotation, Innov. Food Sci. Emerg. Technol. 9 (2008) 320–327.

[49] T.D. Kubritskaya, I.V. Drako, V.N. Sorokina, R.V. Drondina, Use of electrochemical methods to purify the waste water from the production of concentrates in the food industry, Surf. Eng. Appl. Electrochem. 6 (2000) 62–68.

[50] V.I. Il'in, V.A. Kolesnikov, Yu.I. Parshina, Purification of highly concentrated industrial sewage from the porcelain and faience industry by the electric flotation method, Glass Ceram. 59 (7–8) (2002) 242–244.

[51] P. Costaz, J. Miquel, M. Reinbold, Simultaneous electroflotation and disinfection of sewage, Water Res. 17 (3) (1983) 255–262.

[52] I.A. Zolotukhin, V.A. Vasev, A.L. Lukin, Electroflotation purification of the pit waters of the Kuzbass, Khimiya i Tekhnologiya Vody 5 (3) (1983) 252–255.

[53] Y. Fukui, S. Yuu, Removal of colloidal particles in electroflotation, AIChE J. 31 (2) (1985) 201–208.

[54] C. Manohar, V.K. Kelkar, J.V. Yakhmi, Electroflotation of colloids without surfactants, J. Colloid Interface Sci. 89 (1) (1982) 54–60.

[55] E. Sandbank, G. Shelef, A.M. Wachs, Improved electroflotation for the of suspended solids from algal pond effluents removal, Water Res. 8 (1974) 587–592.

[56] J. Ge, J. Qu, P. Lei, H. Liu, New bipolar electrocoagulation–electroflotation process for the treatment of laundry wastewater, Sep. Purif. Technol. 36 (2004) 33–39.

[57] M. Belkacem, M. Khodir, S. Abdelkrim, Treatment characteristics of textile wastewater and removal of heavy metals using the electroflotation technique, Desalination 228 (2008) 245–254.

[58] A.H. Essadki, M. Bennajah, B. Gourich, Ch. Vial, M. Azzi, H. Delmas, Electrocoagulation/electroflotation in an external-loop airlift reactor—application to the decolorization of textile dye wastewater: acase study, Chem. Eng. Process. 47 (2008) 1211–1223.

[59] L. Ben Mansour, I. Ksentini, B. Elleuch, Treatment of wastewaters of paper industry by coagulation–electroflotation, Desalination 208 (2007) 34–41.

[60] L. Ben Mansour, I. Kesentini, Treatment of effluents from cardboard industry by coagulation–electroflotation, J. Hazard. Mater. 153 (2008) 1067–1070.

[61] M.F. Prokop'eva, V.N. Tkacheva, E.Yu. Kirshina, A gas chromographic investigation of the composition of spent cooling lubricants and the products formed during their electroflotation-coagulation treatment, Khimiya i Tekhnologia Vody 10 (4) (1988) 335–337.

[62] O.R. Shendrik, E.E. Andreeva, M.I. Ponomareva, I.B. Ivanenko, Electroflotation treatment of fat-containing solutions, Khimiya i Tekhnologiya Vody 15 (1) (1993) 54–56.

[63] M.N. Rabilizirov, A.M. Gol'man, Treatment of diary waste waters by foam electroflotation-separation, Khimiya i Tekhnologiya Vody 8 (4) (1986) 87–88.

[64] Q. Zuo, X. Chena, W. Li, G. Chen, Combined electrocoagulation and electroflotation for removal of fluoride from drinking water, J. Hazard. Mater. 159 (2008) 452–457.

[65] A.H. Essadki, B. Gourich, M. Azzi, Ch. Vial, H. Delmas, Kinetic study of defluoridation of drinking water by electrocoagulation/electroflotation in a stirred tank reactor and in an external-loop airlift reactor, Chem. Eng. J. 164 (2010) 106–114.

[66] A.H. Essadkia, B. Gourich, Ch. Vial, H. Delmas, M. Bennajah, Defluoridation of drinking water by electrocoagulation/electroflotation in a stirred tank reactor with a comparative performance to an external-loop airlift reactor, J. Hazard. Mater. 168 (2009) 1325–1333.

[67] C.T. Wang, W.L. Chou, Y.M. Kuoa, Removal of COD from laundry wastewater by electro-coagulation/electroflotation, J. Hazard. Mater. 164 (2009) 81–86.

[68] Catherine Ricordel, André Darchen, Dimiter Hadjiev, Electrocoagulation–electroflotation as a surface water treatment for industrial uses, Sep. Purif. Technol. 74 (2010) 342–347.

[69] N.P. Brandon, G.H. Kelsall, Growth kinetics of bubbles electrogenerated at microelectrodes, J. Appl. Electrochem. 15 (4) (1985) 475–484.

[70] V.A. Glembotskii, A.A. Mamakov, A.M. Ramanov, V.E. Nenno, in: Proceedings of the 11th International Mineral Processing Congress, Cagliari, (1975), pp. 562–581.

[71] C. Llerena, J.C.K. Ho, D.L. Piron, Effect of pH on electroflotation of sphalerite, Chem. Eng. Commun. 155 (1996) 217–228.

[72] A.V. Nguyen, H.J. Schulze, Colloidal science of flotation, in: A.T. Hubbard (Ed.), Surfactant Science Series, Marcel Dekker, Inc., New York, Basel, 2004.

[73] V. Sarrot, Z. Huang, D. Legendre, P. Guiraud, Experimental determination of particles capture efficiency in flotation, Chem. Eng. Sci. 62 (24) (2007) 7359–7369.

[74] L.J.J. Janssen, J.G. Hoogland, The effect of electrolytically evolved gas on the thickness of the diffusion layer, Electrochim. Acta. 15 (1970) 1013–1023.

[75] D. Landolt, R. Acosta, R.H. Muller, C.W. Tobais, An optical study of cathodic hydrogen evolution in high rate electrolysis, J. Electrochem. Soc. 117 (6) (1970) 839–845.

[76] P.J. Sides, Phenomena and effects of electrolytic gas evolution, in: R.E. White, J.O'M. Bockris, B.E. Conway (Eds.), Modern Aspects of Electrochemistry, Plenum Press, New York, NY, 1986, pp. 303–354.

[77] L. Ben Mansour, S. Chalbi, I. Kesentini, Experimental study of hydrodynamic and bubble size distributions in electroflotation process, Indian J. Chem. Technol. 14 (3) (2007) 253–257.

[78] J. Venczel, Über den gasblasen bei elektrochemischen prozessen, Electrochim. Acta. 15 (12) (1970) 1909–1920.

[79] V.A. Glembotskii, A.A. Mamakov, V.N. Sorokina, Size of gas bubbles formed under electroflotation conditions, Elektronnaya Obrabotka Materialov. 5 (1973) 66–68.

[80] D.R. Ketkar, R. Mallikarjunan, S. Venkatachalam, Size determination of electrogenerated gas bubbles, J. Electrochem. Soc. India 37 (4) (1988) 313–318.

[81] D. Lumanauw, Hydrogen Bubble Characterization in Alkaline Water Electrolysis. MSc Thesis, Department of Metallurgy and Materials Science, University of Toronto, Canada, 2000.

[82] D.R. Ketkar, R. Mallikarjunan, S. Venkatachalam, Electroflotation of quartz fines, Int. J. Miner. Proc. 31 (1991) 127–138.

[83] V.I. Il'in, O.N. Sedashova, An electroflotation method and plant for removing oil products from effluents, Chem. Petrol. Eng. 35 (1999) 480–481.

[84] A. Khelifa, S. Moulay, A.W. Naceur, Treatment of metal finishing effluents by the electroflotation technique, Desalination 181 (2005) 27–33.

[85] L.S. Tsai, B. Hernlem, C.C. Huxsoll, Disinfection and solids removal of poultry chiller water by electroflotation, J. Food Sci. 67 (2002) 2160–2164.

[86] Y.G. Choi, H.S. Kim, Y.H. Park, S.H. Jeong, D.H. Son, Y.K. Oh, I.T. Yeom, Improvement of the thickening and dewatering characteristics of activated sludge by electroflotation (EF), Water Sci. Technol. 52 (10–11) (2005) 219–226.

[87] X.M. Chen, G. Chen, P.L. Yue, Investigation on the electrolysis voltage of electrocoagulation, Chem. Eng. Sci. 57 (2002b) 2449–2455.

[88] F. Shen, X.M. Chen, P. Gao, G. Chen, Electrochemical removal of fluoride ions from industrial waste water, Chem. Eng. Sci. 58 (2003) 987–993.

[89] X. Chen, G.H. Chen, P.L. Yue, Separation of pollutants from restaurant wastewater by electrocoagulation, Sep. Purif. Technol. 19 (2000) 65–76.

[90] S.E. Burns, S. Yiacoumi, C. Tsouris, Microbubble generation for environmental and industrial separations, Sep. Purif. Technol. 11 (1997) 221–232.

[91] G. Chen, X. Chen, P.L. Yue, Electrocoagulation and electroflotation of restaurant wastewater, J. Environ. Eng. 126 (9) (2000) 858–863.

[92] H.B. Beer, Electrode and coating therefore, USA Patent 3 (632) (1972) 498.

[93] D.M. Novak, B.V. Tilak, B.E. Conway, Fundamental and applied aspects of anodic chlorine production, in: J.O. Bockris, B.E. Conway, R.E. White (Eds.), Modern Aspects of Electrochemistry, vol. 14, Plenum Press, New York, NY, 1982, pp. 195–318.

[94] F. Hine, M. Yasuda, T. Noda, T. Yoshida, J. Okuda, Electrochemical behavior of the oxide-coated metal anodes, J. Electrochem. Soc. 126 (1979) 1439–1445.

[95] V.A. Alves, L.A.D. Silva, E.D. Oliveira, J.F.C. Boodts, Investigation under conditions of accelerated anodic corrosion of the effect of TiO2 substitution by CeO2 on the stability of Ir-based ceramic coatings, Mater. Sci. Forum 289 (1998) 655–666. 292.

[96] X. Chen, G.H. Chen, P.L. Yue, Stable Ti/IrOx–Sb2 O5–SnO2 anode for O2 evolution with low Ir content, J. Phys. Chem. B. 105 (2001) 4623–4628.

[97] G. Chen, X. Chen, P.L. Yue, Electrochemical behavior of stable Ti/IrOx–Sb2 O5–SnO2 anodes for oxygen evolution, J. Phys. Chem. B. 106 (17) (2002) 4364–4369.

[98] X. Chen, G. Chen, P.L. Yue, A novel electrode system for electro-flotation of wastewaters, Environ. Sci. Technol. 36 (4) (2002) 778–783.

[99] V.E. Cenkin, A.N. Belevstev, Electrochemical treatment of industrial wastewater, Eff. Water Treat. J. 25 (7) (1985) 243–249.

[100] A.B. Paul, Proceedings of the 22nd WEDC Conference on Water Quality and Supply, New Delhi, India, 1996, p. 286

[101] S.H. Lin, C.F. Peng, Treatment of textile waste-water by electrochemical method, Water Res. 28 (2) (1994) 277–282.

[102] S.H. Lin, C.F. Peng, Continuous treatment of textile wastewater by combined coagulation, electrochemical oxidation and activated sludge, Water Res. 30 (1996) 587–592.

[103] S.H. Lin, M.L. Chen, Treatment of textile waste-water by chemical methods for reuse, Water Res. 31 (4) (1997) 868–876.

[104] L.J. Gao, Y.F. Cheng, Treatment of printing and dyeing wastewater using pulsed high voltage electrocoagulation flocculation method, Environ. Pollut. Control 14 (5) (1992) 10–13 (in Chinese).

[105] S.H. Lin, C.S. Lin, Reclamation of wastewater effluent from a chemical fiber plant, Desalination 120 (1998) 185–195.

[106] T.Ya. Pazenko, T.I. Khalturina, A.F. Kolova, I.S. Rubailo, Electrocoagulation treatment of oil-containing wastewaters, J. Appl. USSR 58 (1985) 2383–2387.

[107] J. Szynkarczuk, J. Kan, T.A.T. Hassan, J.C. Donini, Clays, Clay Miner. 42 (1994) 667.

[108] R.R. Renk, Electrocoagulation of tar sand and oil shale wastewaters, Energy Prog. 8 (1988) 205–208.

[109] M.F. Pouet, A. Grasmick, Urban wastewater treatment by electrocoagulation and flotation, Water Sci. Technol. 31 (1995) 275–283.

[110] T.R. Demmin, K.D. Uhrich, Improving Carpet Wastewater Treatment, American Dyestuff Reports 77 (1988) 13–18. 32.

[111] G.V. Sleptsov, A.I. Gladkii, E.Y. Sokol, S.P. Novikova, Electrocoagulation treatment of oil emulsion wastewaters of industrial enterprises, Elektronnaya Obrabotka Materialov. 6 (1987) 69–72.

[112] U.B. Ogutveren, S. Koparal, Electrocoagulation for oil–water emulsion treatment, J. Environ. Sci. Health A. 32 (9–10) (1997) 2507–2520.

[113] N.S. Abuzaid, Z. Al-Hamouz, A.A. Bukhari, M.H. Essa, Electrochemical treatment of nitrite using stainless steel electrodes, Water Air Soil Pollut. 109 (1999) 429–442.

[114] U.B. Ogutveren, S. Koparal, Electrochemical treatment of water containing dye-stuffs: anodic oxidation of congo red and xiron blau 2RHD, Int. J. Environ. Stud. 42 (1992) 41–52.

[115] F. Akbal, S. Camci, Copper, chromium and nickel removal from metal plating wastewater by electrocoagulation, Desalination 269 (2011) 214–222.

[116] I. Heidmann, W. Calmano, Removal of Zn(II), Cu(II), Ni(II), Ag(I) and Cr(VI) present in aqueous solutions by aluminium electrocoagulation, J. Hazard. Mater. 152 (2008) 934–941.

[117] W.L. Chou, C.T. Wang, Kai-Yu Huang, Effect of operating parameters on indium (III) ion removal by iron electrocoagulation and evaluation of specific energy consumption, J. Hazard. Mater. 167 (2009) 467–474.

[118] A. Shafaei, M. Rezayee, M. Arami, M. Nikazar, Removal of Mn2+ ions from synthetic wastewater by electrocoagulation process, Desalination 260 (2010) 23–28.

[119] I. Heidmann, W. Calmano, Removal of Ni, Cu and Cr from a galvanic wastewater in an electrocoagulation system with Fe- and Al-electrodes, Sep. Purif. Technol. 71 (2010) 308–314.

[120] L.A. Kul'skii, P.P. Strokach, V.A. Slipchenko, E.I. Saigak, Water Purification by Electrocoagulation, Budivel'nik, Kiev, 1978.

[121] I. Ali, T.A. Khan, M. Asim, Removal of arsenic from water by electrocoagulation and electrodialysis techniques, Sep. Purif. Rev. 40 (2011) 25–42.

[122] E. Lacasa, P. Cañizares, C. Sáez, F.J. Fernández, M.A. Rodrigo, Removal of arsenic by iron and aluminium electrochemically assisted coagulation, Sep. Purif. Technol. 79 (2011) 15–19.

[123] M. Kobya, U. Gebologlu, F. Ulu, S. Oncela, E. Demirbas, Removal of arsenic from drinking water by the electrocoagulation using Fe and Al, Electrochimica Acta. 56 (2011) 5060–5070.

[124] I. Ali, T.A. Khan, M. Asim, Removal of arsenate from groundwater by electrocoagulation method, Environ. Sci. Pollut. Res. 19 (2012) 1668–1676. http://dx.doi.org/10.1007/s11356-011-0681-3.

[125] Xu Zhao, Baofeng Zhang, Huijuan Liu*, Jiuhui Qu, Removal of arsenite by simultaneous electro-oxidation and electro-coagulation process, J. Hazard. Mater. 184 (2010) 472–476.

[126] H. Brammer, P. Ravenscroft, Arsenic in groundwater: a threat to sustainable agriculture in South and South-east Asia, Environ. Int. 35 (2009) 647–654.

[127] S. Gao, J. Yang, J. Tian, F. Ma, G. Tu, M. Du, Electro-coagulation–flotation process for algae removal, J. Hazard. Mater. 177 (1–3) (2010) 336–343.

[128] P. Holt, G. Barton, C. Mitchell, Electrocoagulation as a wastewater treatment. The Third Annual Australian Environmental Engineering Research Event, Castlemaine, Victoria, 1999.

[129] G.H. Chen, Electrochemical technologies in wastewater treatment, Sep. Purif. Technol. 38 (2004) 11–41.

[130] H.M. Wong, C. Shang, Y.K. Cheung, G. Chen, Chloride assisted electrochemical disinfection, in: Proceedings of the Eighth Mainland-Taiwan Environmental Protection Conference Tsin Chu, Taiwan, 2002.

[131] A. Pinisakul, C. Polprasert, P. Parkpian, J. Satayavivad, Arsenic removal efficiency and mechanisms by electro-chemical precipitation process, Water Sci. Technol. 46 (2002) 247–254.

[132] Arsenic removal by electrocoagulation using combined Al–Fe electrode system and characterization of products

[133] I. Arslan-Alaton, I. Kabdasl, Y. Sahin, Effect of Operating Parameters on the Electrocoagulation of Simulated Acid Dye bath Effluent, The Open Environmental & Biological Monitoring Journal 1 (2008) 1–7.

[134] A.T. Kuhn, Electrolytic decomposition of cyanides, phenols and thiocyanates in effluents streams – a literature review, J. Appl. Chem. Biotechnol. 21 (1971) 29–34.

[135] M.O. Azzam, Y. Tahboub, M. Al-Tarazi, Effect of counter electrode material on the anodic destruction of 4-Cl phenol solution, Trans. IChemE B. 77 (1999) 219–226.

[136] M. Gattrell, D.W. Kirk, The electrochemical oxidation of aqueous phenol at a glassy carbon electrode, Can. J. Chem. Eng. 68 (1990) 997–1003.

[137] O.J. Murphy, G.D. Hitchens, L. Kaba, C.E. Verostko, Direct electrochemical oxidation of organics for wastewater treatment, Water Res. 26 (1992) 443–451.

[138] L. Szpyrkowicz, J. Naumczyk, F. Zilio-Grandi, Application of electrochemical processes for tannery wastewater treatment, Toxicol. Environ. Chem. 44 (1994) 189–202.

[139] G. Rajalo, T. Petrovskaya, Selective electrochemicaloxidation of sulphides in tannery wastewater, Environ. Technol. 17 (1996) 605–612.

[140] N.N. Rao, K.M. Somasekhar, S.N. Kaul, L. Szpyrkowicz, Electrochemical oxidation of tannery, J. Chem. Technol. Biotechnol. 76 (2001) 1124–1131.

[141] J.L. Boudenne, O. Cerclier, J. Galea, E.V. Vlist, Electrochemical oxidation of aqueous phenol at a carbon black slurry electrode, Appl. Catal. A: Gen. 143 (1996) 185–202.

[142] O. Cerclier, Performance of carbon blackslurry electrodes for 4-chlorophenol oxidation, Water Res. 33 (1999) 494–504.

[143] A.M. Polcaro, S. Palmas, Electrochemical oxidation of chlorophenols, Ind. Eng. Chem. Res. 36 (1997) 1791–1798.

[144] J. Manriquez, J.L. Bravo, S. Gutierrez-Granados, S.S. Succar, C. Bied-Charreton, A.A. Ordaz, F. Bedioui, Electrocatalysis of the oxidation of alcohol and phenol derivative pollutants at vitreous carbon electrode coated by nickel macrocyclic complex-based films, Anal. Chim. Acta. 378 (1999) 159–168.

[145] C.S. Hofseth, T.W. Chapman, Electrochemical destruction of dilute cyanide by copper-catalyzed oxidation in a flowthrough porous electrode, J. Electrochem. Soc. 146 (1999) 199–207.

[146] X. Chen, G. Chen, P.L. Yue, Anodic oxidation of dyes at novel Ti/B-diamond electrode, Chem. Eng. Sci. 58 (2003) 995–1001.

[147] I. Troster, L. Schafer, M. Fryda, Recent developments in production and application of DiaChem-electrodes for wastewater treatment, New Diam. Front. C. Technol. 12 (2) (2002) 89–97.

[148] C. Comninellis, E. Plattner, Electrochemical wastewater treatment, Chimia. 42 (7–8) (1988) 250–252.

[149] Y.M. Awad, N.S. Abuzaid, Electrochemical treatment of phenolic wastewater: efficiency, design considerations and economic evaluation, J. Environ. Sci. Health A. 32 (1997) 1393–1414.

[150] N. Kannan, S.N. Sivadurai, L.J. Berchmans, R. Vijayavalli, Removal of phenolic compounds by electrooxidation method, J. Environ. Sci. Health A. 30 (1995) 2185–2203.

[151] I.H. Yeo, D.C. Johnson, Electrocatalysis of anodic oxygentransfer reactions: effect of groups IIIA and VA metal oxides in electrodeposited -lead dioxide electrodes in acidic media, J. Electrochem. Soc. 134 (1987) 1973–1977.

[152] J. Feng, D.C. Johnson, Electrocatalysis of anodic oxygentransfer reaction: titanium substrates for pure and doped lead dioxide films, J. Electrochem. Soc. 138 (1991) 3329–3337.

[153] J. Feng, L.L. Houk, D.C. Johnson, Electrocatalysis of anodic oxygen-transfer reactions: the electrochemical incineration of benzoquinone, J. Electrochem. Soc. 142 (1995) 3626–3631.

[154] A.M. Polcaro, S. Palmas, F. Renoldi, M. Mascia, On the performance of Ti/SnO_2 and Ti/PbO_2 anodes in electrochemical degradation of 2-chlorophenol for wastewater treatment, J. Appl. Electrochem. 29 (1999) 147–151.

[155] D.W. Kirk, H. Sharifian, F.R. Foulkes, Anodic oxidation of aniline for waste water treatment, J. Appl. Electrochem. 15 (1985) 285–292.

[156] C.A. Vincent, D.G.C. Weston, Preparation and properties of semiconducting polycrystalline tin oxide, J. Electrochem. Soc. 119 (1972) 518–521.

[157] J.A. Aboaf, V.C. Marcotte, Chemical composition and electrical properties of tin oxide films prepared by vapor deposition, J. Electrochem. Soc. 120 (1973) 701–702.

[158] Z.M. Jarzebski, J.P. Marton, Physical properties of SnO2 materials. I. Preparation and defect structure, J. Electrochem. Soc. 123 (1976) 199C–205C.

[159] Y.S. Hsu, S.K. Ghandhi, The preparation and properties of arsenic-doped tin oxide films, J. Electrochem. Soc. 126 (1979) 1434–1435.

[160] Y.S. Hsu, S.K. Ghandhi, The effect of phosphorus doping on tin oxide films made by the oxidation of phosphine and tetramethyltin. I. Growth and etching properties, J. Electrochem. Soc. 127 (1980) 1592–1595.

[161] Y.S. Hsu, S.K. Ghandhi, The effect of phosphorus doping on tin oxide films made by the oxidation of phosphine and tetramethyltin. II. Electrical properties, J. Electrochem. Soc. 127 (1980) 1595–1599.

[162] R. Kotz, S. Stucki, B. Carcer, Electrochemical waste water treatment using high overvoltage anodes. Part I. Physical and electrochemical properties of SnO2 anodes, J. Appl. Electrochem. 21 (1991) 14–20.

[163] Ch. Comninellis, Electrochemical treatment of wastewater containing phenol, Trans. IChemE B. 70 (1992) 219–224.

[164] A. Nanthakumar, N.R. Armstrong, Studies in physical and theoretical chemistry, in: H.O. Finklea (Ed.), Semiconductor Electrodes, vol. 55, Elsevier, Amsterdam, 1988, p. 203.

[165] F. Grimm, D. Bessarabov, W. Maier, S. Storck, R.D. Sanderson, Sol–gel film-preparation of novel electrodes for the electrocatalytic oxidation of organic pollutants in water, Desalination 115 (1998) 295–302.

[166] R. Cossu, A.M. Polcaro, M.C. Lavagnolo, M. Mascia, S. Palmas, F. Renoldi, Electrochemical treatment of landfill leachate: oxidation at Ti/PbO_2 and Ti/SnO_2 anodes, Environ. Sci. Technol. 32 (1998) 3570–3573.

[167] B. Correa-Lozano, Ch. Comninellis, A. Debattisti, Service life of Ti/SnO_2–Sb2O_5 anodes, J. Appl. Electrochem. 28 (8) (1997) 970–974.

[168] J.M. Kesselman, O. Weres, N.S. Lewis, M.R. Hoffmann, Electrochemical production of hydroxyl radical at polycrystalline Nb-doped TiO_2 electrodes and estimation of the partitioning between hydroxyl radical and direct hole oxidation pathways, J. Phys. Chem. B 101 (1997) 2637–2643.

[169] J. Naumczyk, L. Szpyrkowicz, F.Z. Grandi, Electrochemical treatment of textile wastewater, Water Sci. Technol. 34 (11) (1996) 17–24.

[170] A.G. Vlyssides, C.J. Israilides, Detoxification of tannery waste liquors with an electrolysis system, Environ. Pollut. 97 (1–2) (1997) 147–152.

[171] A.G. Vlyssides, C.J. Israilides, M. Loizidou, G. Karvouni, V. Mourafeti, Electrochemical treatment of vinasse from beet molasses, Water Sci. Technol. 36 (2–3) (1997) 271–278.

[172] L.C. Chiang, J.E. Chang, T.C. Wen, Electrochemical oxidation process for the treatment of coke-plant wastewater, J. Environ. Sci. Health A. 30 (1995) 753–771.

[173] A.G. Vlyssides, C.J. Israilides, Electrochemical oxidation of a textile dye and finishing wastewater using a Pt/Ti electrode, J. Environ. Sci. Health A. 33 (1998) 847–862.

[174] E. Brillas, R.M. Bastida, E. Llosa, Electrochemical destruction of aniline and 4-chloroaniline for wastewater treatment using a carbon-PTFE O_2-fed cathode, J. Electrochem. Soc. 142 (1995) 1733–1741.

[175] E. Brillias, E. Mur, J. Casado, Iron(II) catalysis of the mineralization of aniline using a carbon-PTFE O_2-fed cathode, J. Electrochem. Soc. 143 (1996) L49–L53.

[176] E. Brillas, R. Sauleda, J. Casado, Peroxi-coagulation of aniline in acidic medium using an oxygen diffusion cathode, J. Electrochem. Soc. 144 (1997) 2374–2379.

[177] E. Brillas, R. Sauleda, J. Casado, Degradation of 4-chlorophenol by anodic oxidation, electro-Fenton, photoelectro-Fenton, and peroxi-coagulation processes, J. Electrochem. Soc. 145 (1998) 759–765.

[178] W. El-Shal, H. Khordagui, O. El-Sebaie, F. El-Sharkawi, G.H. Sedahmed, Electrochemical generation of ozone for water treatment using a cell operating under natural convection, Desalination 99 (1991) 149–157.

[179] J.C. Farmer, F.T. Wang, R.A. Hawley-Fedder, P.R. Lewis, L.J. Summers, L. Foiles, Electrochemical treatment of mixed and hazardous wastes: oxidation of ethylene glycol and benzene by silver(II), J. Electrochem. Soc. 139 (1992) 654–662.

[180] F. Bringmann, K. Ebert, U. Galla, H. Schmieder, Electrochemical mediators for total oxidation of chlorinated hydrocarbons: formation kinetics of Ag(II), Co(III), and Ce(IV), J. Appl. Electrochem. 25 (1995) 846–851.

[181] V. Cocheci, C. Radovan, G.A. Ciorba, I. Vlaiciu, Mediate electrochemical wastewater treatment, Revue. Roumaine de Chimie. 40 (1995) 615–619.

[182] A. Paire, D. Espinoux, M. Masson, M. Lecomte, Silver(II) mediated electrochemical treatment of selected organics: hydrocarbon destruction mechanism, Radiochim. Acta. 78 (1997) 137–143.

7 Wastewater Treatment by Biological Methods

CHAPTER OUTLINE

Environmental Water, http://dx.doi.org/10.1016/B978-0-444-59399-3.00007-6

7.1 Introduction

Water whose quality has been adversely affected by anthropogenic influence constitutes a major issue of concern. Wastewater comprises liquid waste discharged by domestic residences, commercial properties, industries, and agricultural activities. This usually encompasses a wide range of potential contaminants. Both organic and mineral contaminants present in the industrial wastewater exhibit toxic effects on the microbiological agents of wastewater treatment plants and often resist biological attack and mineralization [1]. Wastewater treatment is achieved via a large array of methods and processes among which biological methods are of considerable interest. The main contaminants present in the waste water are nitrogen (organic or inorganic) in the form of urea, ammonia, nitrites, and nitrates; phosphorus, proteins, drugs, pharmaceuticals, sea-salt, cyanides, thiocyanates, thiosulfates; pathogens such as bacteria, viruses, and parasitic worms; inorganic particles such as sand, grit, metal particles, and ceramics; gases such as hydrogen sulfide, carbon dioxide, methane; emulsions such as paints, adhesives, hair colorants, and emulsified oils. Approximations suggest that water will become one of the scarcest resources in the near future due to massive worldwide increase in the human population [2]. The collection of wastewater has been encouraged from the ancient times. However, its treatment is a bit new development which was made during late 1800s and early 1900s [3]. Many biological methods by different workers have been developed from time to time to carry out the treatment of wastewater. Some of the methods are quite good and can be therefore applied for water reuse. All the biological methods of wastewater treatment work on the fact that bacterial cells feed on organic materials in the wastewater and thereby reduce its BOD (biological oxygen demand) content. Therefore, the purpose of biological treatment is the reduction of BOD. Most often wastewater with BOD higher than 200 mg/L enters the treatment plant. However, primary settling reduces it to about 150 mg/L by the time it enters the biological component of the system. The BOD content of the water leaving the system should not be higher than 20–30 mg/L, so that after dilution in the nearby receiving water body (river, lake), the BOD is less than 2–3 mg/L. Simple bacteria utilize the organic material present in the wastewater as food and, therefore, convert the organic material into cellular mass which gets precipitated at the bottom of a settling tank or is retained as slime on solid surfaces or vegetation in the system. This whole process makes the exiting water much cleaner than the water entering the system. Adequate supply of oxygen forms a key factor is the operation of any biological system since bacterial cells need not only organic material as food but also oxygen to breathe (aerobic methods).

Biological treatment of the wastewater involves operation at ambient temperature. Energy is saved as it involves no heating or cooling of the water. Wastewater treatment operations are located outdoor since they take much space. Cells come in a mix of many types, and accommodation to a temperature change is simply accomplished by self adaptation of the cell population. A change in composition of the organic material leads to a spontaneous change in cell population, with the types best suited to digest the new material growing in larger numbers than other cell types. There is a great variation among the methods used for the treatment of wastewater all over the world. Treatment and discharge systems in developed nations not only differ between countries but also among rural and urban users [4]. The authors are of the opinion that in developed countries, the most common wastewater treatment methods are centralized aerobic wastewater treatment plants and lagoons. However, in developed countries, there is still a fraction of population without adequate wastewater treatment facilities. In certain cases, the infrastructure is inadequate; and even in areas with a high degree of wastewater treatment, certain pathogens and chemicals may still make their entry into the environment [5].

7.2 Stages of Biological Treatment of Wastewater

Wastewater treatment is achieved via preliminary treatment, primary treatment, secondary treatment, and advanced or tertiary treatment (Figure 7-1). All these processes involve the active participation of biological agents. These processes are briefly described as follows:

7.2.1 Preliminary Treatment

Preliminary treatment forms the first step in any water treatment system. This treatment simply removes large floating objects that may clog pumps, grit, and small pipes [6]. The removal of these materials enhances the maintenance and operation of subsequent treatment units. The velocity of water through the grit chambers is maintained sufficiently high, or air is used, to prevent the settling of most organic solids. Most often, flow measuring devices like standing wave flumes are always included at the preliminary treatment stage.

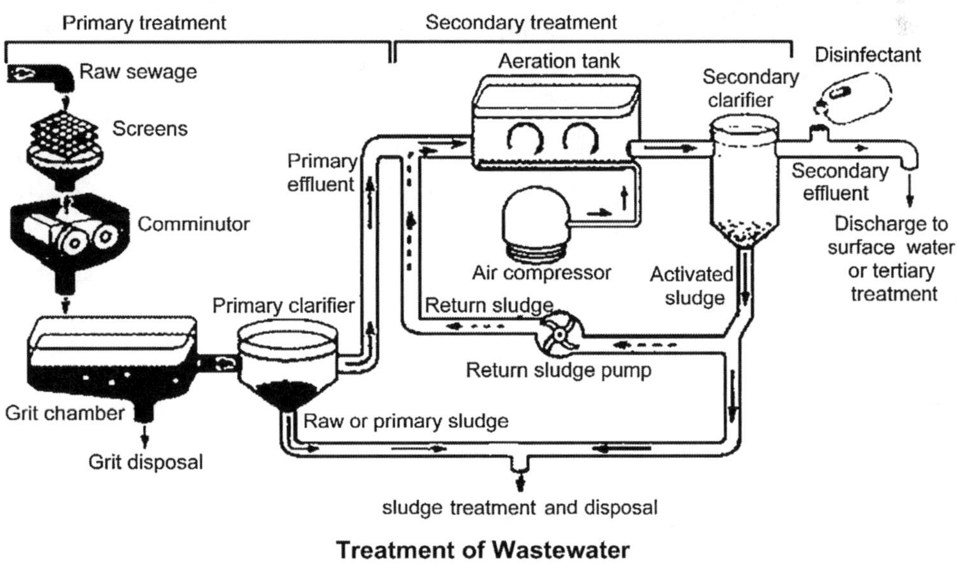

FIGURE 7-1 A general overview of biological treatment of wastewater.

7.2.2 Primary Treatment

The second step in wastewater treatment is the primary treatment. This step separates suspended solids (SS) and grease from wastewater [6]. Primary and secondary stages may be combined into one basic operation in certain treatment plants. After the grit is removed, wastewater still contains some dissolved organic and inorganic constituents along with SS. After wastewater enters a sedimentation tank, it slows down. The SS in the slowed down wastewater gradually sink to the bottom, as primary sludge which can then be removed from the tank by various methods [7]. The primary treatment approximately removes 25–50% of the incoming biochemical oxygen demand, 65% of the oil and grease, and 50–70% of the total SS. Moreover, some organic phosphorus, organic nitrogen, and some heavy metals are also removed during primary sedimentation but this process does not affect colloidal and dissolved constituents.

7.2.3 Secondary Treatment

In this stage of biological treatment of wastewater, dissolved organic matter is removed and this process accounts for the removal of almost 90 % of the organic matter in wastewater. Herein, sewage microorganisms are cultivated and added to the wastewater. Organic matter from sewage is absorbed by these microorganisms as food and therefore such organic matter gets removed via circulation [6]. Attached growth processes, suspended growth processes, and lagoon systems are the three most common conventional methods used to carry out secondary treatment [8]. In attached growth processes, microbes are allowed to grow on surfaces such as stone or plastic media. Wastewater flows over the media and also provides oxygen. The growth processes greatly help in the removal of biodegradable organic material [7]. Suspended growth processes are designed in such a way that they remove biodegradable organic material and organic nitrogen-containing material.

7.2.4 Advanced or Tertiary Treatment

Tertiary treatment is the additional treatment required to remove suspended and dissolved substances which remain after conventional secondary treatment. This may be achieved by using various physical, chemical, or biological treatment processes [7]. Filtration, removal of ammonia, and other pollutants and disinfection to destroy pathogens are the main processes included in tertiary treatment [9].

7.2.5 Disinfection

Most often inadequately treated wastewaters contain different species of pathogens. Disinfection is the process that usually kills or sometimes deactivates these harmful pathogens. The most commonly used disinfecting agent is chlorine, although ozone and ultraviolet (UV) radiation also find frequent uses for wastewater effluent disinfection. Chlorine is applied to wastewater as a gas, liquid, or in a solid form and kills microorganisms by destroying their cellular contents. Free or unused chlorine that sometimes remains in water, even at low concentrations, can be hazardous to aquatic life [9] and hence, it is mandatory to remove free chlorine by dechlorination in order to protect fish and aquatic life. Ozone used for disinfection is very effective for killing viruses and bacteria. The main advantage associated with the use of ozone as a disinfectant is that it decomposes back to oxygen rapidly without leaving harmful by-products. However, the high energy costs needed for the use of ozone make its application a bit tedious [10]. Disinfection with UV radiation is a physical treatment process and leaves no chemical traces. However, when UV is used at low doses, the treated pathogens can repair and reverse its destructive effects [11]. Schaar et al. [12] installed a pilot-scale ozonation plant which was for micropollutant removal during biological wastewater treatment. Removal of most of the micropollutants, especially the compounds that were not degraded in the previous biological process, e.g. carbamazepine and diclofenac, was quite increased by the application of 0.6 g O_3. The authors were conclusive of the fact that ozonation of wastewater treatment plant effluent is a promising technology to further decrease emissions of micropollutants during treatment processes.

7.3 Methods of Biological Wastewater Treatment

7.3.1 Oxidation Ponds

Oxidation ponds are effective, low cost, and simple technology for reducing the BOD of wastewater before it is discharged to an aquatic ecosystem (Figure 7-2). It consists of ring or oval shaped channel equipped with mechanical aeration devices. Screened wastewater entering the pond is aerated by mechanical devices which circulates at about 0.25–0.35/ ms. Oxidation ponds typically operate in an extended aeration mode with long detention and solids retention times [13]. Kotsou et al. [14] herein described an aerobic biological

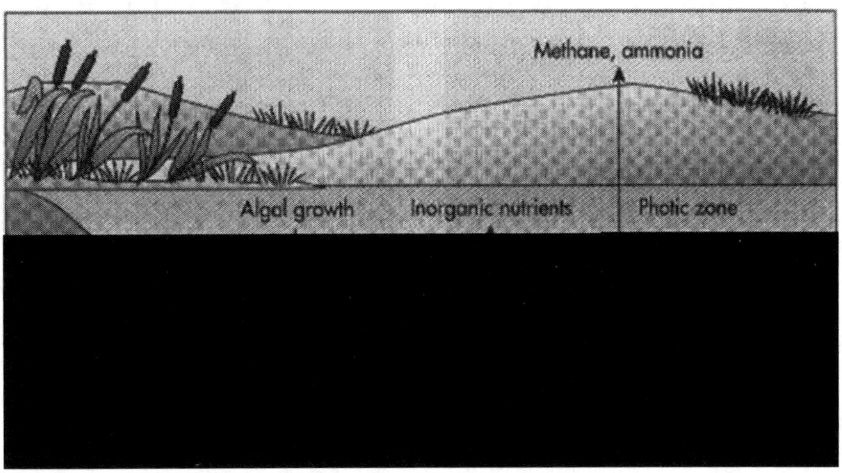

FIGURE 7-2 Basics of an oxidation pond. For color version of this figure, the reader is referred to the online version of this book.

method using an *Aspergillus niger* strain in a bubble column bioreactor in combination with chemical oxidation treatment of table olive processing wastewater. As per the reports by the authors, a quite significant reduction of chemical oxygen demand (COD) by 70% while the total and simple phenolic compounds were decreased by 41 and 85%, respectively, after 2 days of biological treatment. Chemical oxidation step mainly effected the elimination of persistent phenolic compounds during the biological treatment of total phenolic compounds. Moreover, CaO assisted coagulation was found to significantly improve the efficiency of the process. Zhao et al. [15] described a hybrid wastewater treatment process consisting of neutralization, precipitation, modified two-phase anaerobic digestion, and biological aerated filter (BAF) for the successful treatment of yellow ginger processing wastewater with high COD and high sulfate. The inhibition of the methanogens was eliminated to the greatest extent by the modified two-phase anaerobic digestion, which resulted in a high-efficiency COD removal. The effluent from the anaerobic units was treated in the novel BAF which was equipped with special carriers with excellent results. Simultaneous nitrification and denitrification process was held responsible for these results. Baban et al. [16] herein described a method depicting the biological and oxidative treatment of cotton textile dye-bath effluents by fixed and fluidized bed reactors. The methodology utilizes ozone oxidation to assess the combination of biological and chemical oxidation. Approximately 99% color removal was achieved and COD removal efficiencies ranged from 70 to 93%. The overall reported treatment system has economical and operational advantages as compared to treatment options such as advanced oxidation processes (AOPs) using UV, O_3, H_2O_2, or electrocoagulation. Ballesteros et al. [17] herein described the effect of pesticide concentration (expressed as dissolved organic carbon) on combined solar photo-Fenton and biological oxidation treatment. Biodegradation kinetics indicated that the intermediates generated at the higher pesticide concentration caused lower carbon removal rates in spite of the longer photo-Fenton treatment time applied. For the treatment of water with high concentrations of pesticides and to overcome the low biodegradability of photo-Fenton intermediates, it is suggested to mix it with a biodegradable carbon source before biological oxidation.

7.3.2 Anaerobic Ponds

Anaerobic ponds (Figure 7-3) are designed in such a way that the pretreatment of high-strength wastewaters can be achieved efficiently. This is used prior to the aerobic treatment and is often very effective and economical for removing high concentrations of BOD and COD [18]. These ponds have more organic loading relative to the amount of oxygen entering the pond for maintaining anaerobic conditions to the pond surface. Naturally occurring anaerobic bacteria break down the organic content in the effluent, releasing methane and carbon dioxide. A crust may form on the surface and sludge is deposited at the bottom [4]. Fu et al. [19] described textiles wastewater treatment using anoxic filter bed (AFB) and (BWB-BAF). The study indicated that the effluent COD concentration of the AFB decreased with new carriers added and 20.2% was the average removal COD efficiency. In

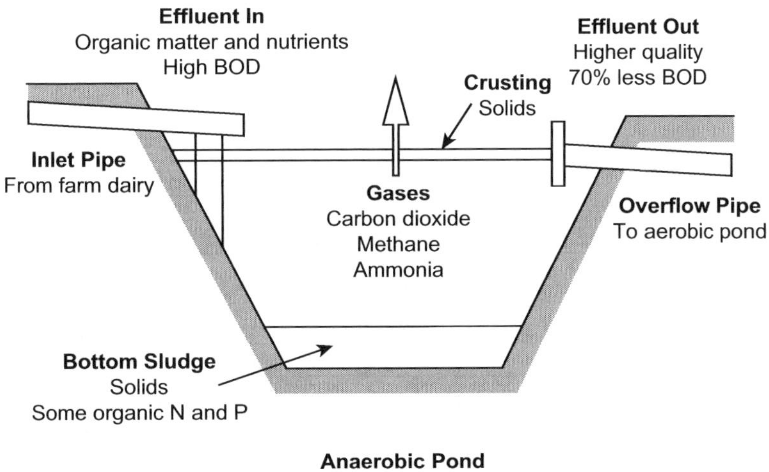

FIGURE 7-3 An overview of an anaerobic pond.

conclusion, this study is a beneficial attempt to utilize the AFB-BWB-O_3-BAF combine process for textile wastewater treatment. Leal et al. [20] herein described the occurrence of xenobiotics in gray water and removal in three biological treatment systems. In this study, 18-selected xenobiotics related to personal care and household chemicals were estimated in gray water from 32 houses and in effluents of three different biological treatment systems (aerobic, anaerobic, and combined anaerobic + aerobic). The authors reported that the removal efficiencies were higher under aerobic conditions than under anaerobic conditions. Chan and et al. [21] herein described a biological treatment of anaerobically digested palm oil mill effluent (POME) using a sequencing batch reactor (SBR). The authors assessed the performance of the SBR by measuring COD, BOD, and total suspended solids (TSS) removal and sludge volume index (SVI). It was observed that the effluent quality remained stable and complied with the discharge limit. Moreover, the sludge showed good settling properties with average SVI of 65. Therefore, it was concluded that the SBR process could complement the anaerobic treatment to produce final treated effluent which meets the discharge limit.

7.3.3 Aerobic Ponds

Aerobic ponds (Figure 7-4) contain suspensions of algae and bacteria that maintain aerobic conditions throughout their depth. Mainly two types of aerobic ponds viz shallow ponds and aerated ponds are used [22]. Souza et al. [23] investigated the treatment of refinery wastewater by AOPs in conjunction with biological activated carbon (BAC) with the purpose to produce water for reuse. BAC filters were found to be effective, reaching average efficiencies of 65% in a sufficiently long period of operation (84 days), while granular activated carbon filters were saturated after 28 days. Cao et al. [24] described internal electrolysis-biological contact oxidation process for the treatment of linear

Aerobic Pond

FIGURE 7-4 Picture of an aerobic pond. For color version of this figure, the reader is referred to the online version of this book.

alkylbenzene sulfonate wastewater. The authors reported the treatment of wastewater collected from a daily chemical plant by the combination processes of Fe/C internal electrolysis and biological contact oxidation. The effluent treated by this process could meet the I-grade criteria specified in Integrated Wastewater Discharge Standard of China (GB 8978-1996). Zhu and Chen [25] reported an efficient method for the reduction of nitrous oxide N_2O and nitric oxide (NO) generation in anaerobic–aerobic processes. The generation of N_2O and NO was reduced by 68.7% and 50.0%, respectively, by the use of waste-activated sludge (AS) alkaline fermentation liquid as the synthetic wastewater-carbon source, compared with the commonly used carbon source in the literature (e.g. acetic acid). However, the removal efficiencies of total phosphorus (TP) and total nitrogen (TN) were improved. It was observed that sludge fermentation liquid caused significant decrease of both nitrite reductase activity to NO reductase activity ratio and NO reductase activity to N_2O reductase activity ratio, which resulted in the lower generation of NO and N_2O. Dhadse et al. [26] reported an eco-friendly, cost-effective, and sustainable method for the treatment of herbal pharmaceutical wastewater using earthworms. As per the reports by authors, treated effluent was free of color and odor with efficient COD/BOD removals in the range of 85.44–94.48% and 89.77–96.26%, respectively, at 2 days hydraulic retention time. Removal of heavy metals was also observed with no sludge production problem. Nutrient-rich vermicasts from the filters were removed and analyzed after the experiments which showed higher manurial value than control in terms of available nitrogen, phosphorus, and potassium. Oliveira et al. [27] used an air microbubble bioreactor with a free self-adapted microbial population for aerobic treatment of winery wastewater. The wastewater COD ranged from 4.0–8.0 kg COD/m^3 and the efficiency of the batch treatment

FIGURE 7-5 Construction of a shallow pond. For color version of this figure, the reader is referred to the online version of this book.

was about 90.0 ± 4.3%, after 6 days of operation. The maximum efficiency obtained was achieved after 15 days of treatment (99%). The authors concluded by reporting that this study will contribute for the implementation of an efficient water use plan, aiming the preservation of the water resource and the reduction of the wastewater production. Beline et al. [28] described a biological aerobic treatment of piggery wastewater to control nitrogen, phosphorus, pathogens, and gas emissions. Four types of treatment processes including a biological reactor were built to remove about 60–70% of the nitrogen content as gas by nitrification/denitrification. The different mechanical separators (screw press, centrifuge decanter) were added which led to concentration of phosphorus in an exportable solid phase. A considerable reduction up to 80% of the phosphorus was achieved. Additionally, gaseous emissions were considerably reduced using this management process as compared to the conventional management (storage + land spreading) including ammonia (up to 68%) and greenhouse gases (55%). Moreover, levels of enteric and pathogenic bacteria were also decreased with the treatment process.

7.3.4 Shallow Ponds

Shallow aerobic ponds (Figure 7-5) usually meet the dissolved oxygen demands by transfer of oxygen between air and water. However, the efficiency of soluble biochemical oxygen demand removal can be as high as 95%, the pond effluent will have to contain a large amount of algae which will contribute to the measured total biochemical oxygen demand of the effluent. Moreover, the suspended algae and microorganisms have to be separated from the pond effluent in order to achieve removal of both soluble and insoluble biochemical oxygen demand [29].

7.3.5 Aerated Ponds

Aerated ponds (Figure 7-6) are similar to oxidation ponds. However, such ponds are deeper and require mechanical aeration devices to transfer oxygen into their deeper portions and facilitate a proper mix of the wastewater and bacteria. Aerated ponds require

FIGURE 7-6 Construction of an aerated pond. For color version of this figure, the reader is referred to the online version of this book.

less area than oxidation ponds which makes them better suited. However, they suffer from the disadvantage that the mechanical aeration devices require maintenance and dissipate energy [30]. Yin et al. [31] described a one-step aerobic contact oxidation process and anoxic/aerobic two-step contact oxidation process for the treatment of high-ammonia nitrogen polluted river water. Stepkin and Stepkina [32] devised an apparatus and established a method for aerobic biological treatment of wastewater and complex organomineral fertilizer. This method recycles raw sludge and active sludge subjected to aerobic biological treatment by producing a complex organomineral fertilizer there from, while simultaneously increasing efficiency and quality of wastewater treatment.

7.3.6 Facultative Ponds

These ponds are generally aerobic but can also operate in a facultative manner. They generally have a depth of about 1.0–1.5 m which is too deep for oxygen to penetrate to the bottom of the pond and thereby an anaerobic zone develops at the bottom of the pond. Solid materials that come along with the wastewater get deposited over the anaerobic sludge near the bottom of the pond. The deposited solid materials undergo anaerobic degradation and release soluble degradable organic material and nutrients which diffuse upwards in the pond [33]. Oxygen is supplied by algal photosynthesis and by diffusion from the air near the top of the pond. During night time, dissolved oxygen is present to only a few centimeters depth but often diffuses deeper during daylight [34].

7.3.7 Trickling Filter

A trickling filter (TF) reduces BOD, pathogens, and nitrogen levels of the wastewater. A bed of porous material (slag, rocks, and plastic media) forms the main working part of the trickling filter (Figure 7-7). The microorganisms present in the wastewater get attached to

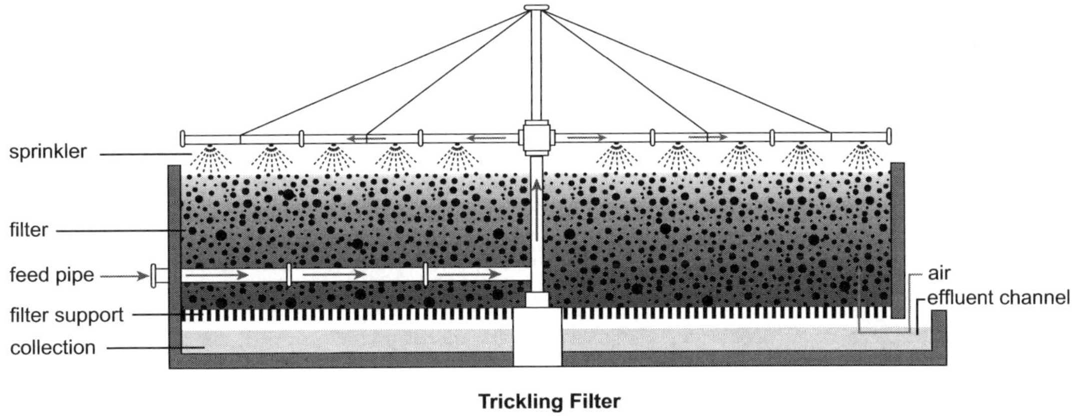

FIGURE 7-7 Construction of a trickling filter. For color version of this figure, the reader is referred to the online version of this book.

the bed which is covered with bacteria. Wastewater first distributes over the media surface where it flows downward as a thin film for aerobic treatment. Finally, it is collected at the bottom through an under drain system. Effluent settles by gravity and biological solids are removed prior to being discharged [35]. Ji et al. [36] described the effectiveness of a multi-media biological aerated filter (MBAF) with clinoptilolite media for the treatment of synthetic wastewater. To the clinoptilolite media of MBAFs in a brick-wall embedded design was added coal ash bioceramsite with supplemental metallic iron. It was observed that the MBAFs with more metallic iron were found to have superior hydraulic and organic loading and higher N and P capacities.

7.3.8 Rotating Biological Contractors

Rotating biological contractors have a number of circular plastic discs mounted on a central shaft (Figure 7-8). These discs remain submerged and continuously rotate in a tank containing the wastewater. The microorganisms rotate in and out of the wastewater by getting attached to the disc. The amount of oxygen required for the conversion of organic matter adsorbed from the liquid is obtained from the air since a certain area of the disc usually rotates out of the liquid [37]. Winery wastewater and effluents produced by various industries such as gold mining and domestic sewage treatment have been successfully treated by rotating biological contractors [38]. Kargi et al. [39] described the biological treatment of saline wastewater in a rotating biodisc contactor (RBC) under different operating conditions like feed COD, A/Q ratio, COD loading, and salt concentrations. The results obtained by the authors were indicative of the fact that an increase in the A/Q ratio increased the percentage COD removal. However, percentage COD removal was found to decrease with feed COD and salt content. Furthermore, COD removal efficiency was found to be more sensitive to changes in A/Q, feed COD, and salt content at low values. Arnaud [40] herein described the treatment of winery wastewater with an anaerobic rotating biological contactor. The authors observed optimal performances with

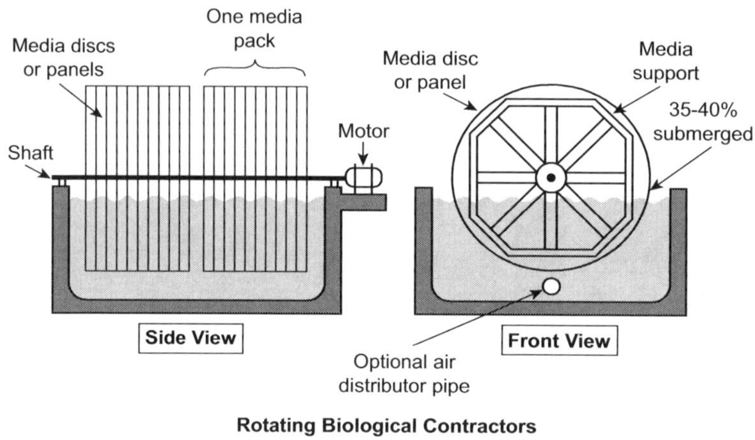

FIGURE 7-8 Construction of a RBC. For color version of this figure, the reader is referred to the online version of this book.

a COD removal close to 80%. However, it was felt that increasing the total surface of the biodisc and controlling temperature to the mesophilic optimal value (37 °C) could lead to better treatment results.

7.3.9 AS Treatment

AS process represents the most commonly and widely applied biological wastewater treatment process all over the world. A diagrammatic view of the working of the treatment process is shown in Figure 7-9. The prime purpose of this process is to remove the soluble

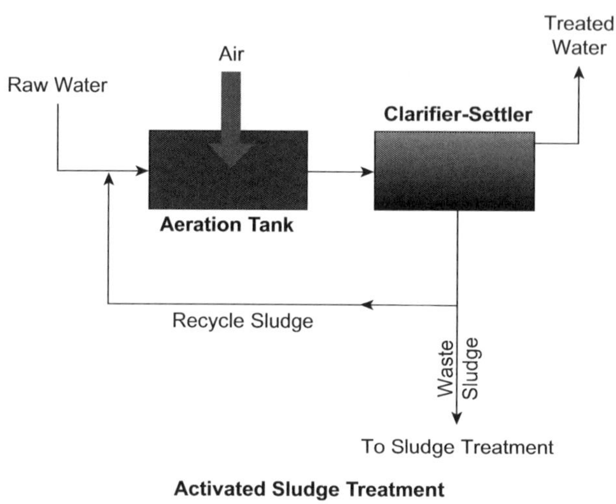

FIGURE 7-9 A pictorial representation of AS treatment. For color version of this figure, the reader is referred to the online version of this book.

biodegradable compounds from wastewater. This process is capable of reducing the soluble substrates in reactors of much smaller volume while producing effluent significantly free of SS [18,41]. The efficiency of removal of pathogenic and indicator microorganisms in these wastewater treatment plants depends on the treatment process, retention time, other biological flora present in AS, pH, temperature, oxygen concentration, and the efficiency of removing SS [4]. Xiang and Jiti [42] demonstrated a method for the complete recycling of wastewater treatment sludge. The beauty of the method lies in the fact that the usually difficult treatment of sludge from wastewater plants becomes quite easy by applying this method. Moreover, water treatment cost and environmental pollution are reduced significantly. Ayano et al. [43] described a method for the treatment of paper manufacturing wastewater using a phosphorus component contained in sludge recovered after AS process involving preparation of a neutral or weakly acidic H_3PO_4-containing solution containing 0.001–1.0 wt % of H_3PO_4 by treating the sludge at 40–100 °C and using the H_3PO_4-containing solution as a microorganism nutrient source for the AS process and/or biological membrane treatment of the treated water. Park et al. [44] described a method for the treatment of wastewater containing dimethyl sulphoxide (DMSO). The authors evaluated the feasibility of the biological treatment of DMSO containing wastewater with AS and also optimized the conditions for Fenton treatment. Despite of the fact that total organic carbon (TOC) and COD removal efficiencies of the Fenton treatment were not satisfactory for most applications, the BOD:COD ratio increased from 0.035 to 0.87. This clearly indicates that it may be a very useful pretreatment method for biological treatment. Wastewater contaminated with 800 mg/L of DMSO was successfully treated by AS without involving Fenton pretreatment, after 20 days of acclimation period. It was observed that most of the sulfur in DMSO oxidized to sulfate, which thereby eradicated the possibility of the production of sulfide-containing noxious intermediates. Moreover, pH was an important operating factor for AS operation. Finally, it was concluded that the wastewater containing DMSO may be successfully treated with this method. Furthermore, this method may significantly reduce the treatment costs as compared to the chemical methods currently under use. Kargi et al. [45] described an effective methodology for the treatment of nitrogen deficient wastewaters by using nitrogen fixing bacteria, *Azotobacter vinelandii* in AS culture. A comparison was made between the COD removal performance of Azotobacter-supplemented AS with Azotobacter-free AS culture for biological treatment of nitrogen deficient synthetic wastewater. Moreover, variable effects on the COD removal performance such as N/COD ratio, sludge age, hydraulic residence time, feed COD concentration, and the COD loading rate were investigated. Finally, the authors concluded that Azotobacter addition to the AS improved the COD removal performance significantly.

Bohdziewicz et al. [46] studied the effectiveness of a hybrid system combining the biological methods of AS (in an SBR) and reverse osmosis for evaluating the effectiveness of the treatment of wastewater coming from meat industry. The authors reported that the wastewater from the Meat Processing Plant Uni-Lang in Wrzosowa underwent sufficient removal of contaminants which consequently could be discharged into receiving water.

Additionally, the wastewater was treated with reverse osmosis for reuse in the production cycle. Li et al. [47] studied the field process performance and microorganism characteristics of enhanced biological phosphorus removal (EBPR) in municipal wastewater treatment plants in China. The authors reported that the three wastewater treatment plants meet the required criteria of phosphorus for discharge ($\leq$1 mg/L), but with a low level of *Accumulibacter* (6.4–3.8%, on average) and a relatively high level of *Competibacter* (3.2–9.1%) in sludge. Zhang et al. [48] herein reported the investigation of the changes in composition and structure of bacterial communities developing on BAC particles and the bacterial functions. The results of the study indicated that the stable bacterial structure did not develop on BAC particles until the 9th month during running time of the BAC filter. *Pseudomonas* sp., *Bacillus* sp., *Nitrospira* sp., and an uncultured bacterium finally dominated the communities. Moreover, the stable bacterial communities played an important role in the removal of ammonia–nitrogen and TOC. Furthermore, employing gas chromatography–mass spectrometry (GC-MS), it was proved that 36 kinds of chemicals in feed water were eliminated and concentrations of five kinds of chemicals decreased. These chemicals served as nutrients for the dominant bacteria. Zhang et al. [49] herein described an improved strategy for the enhanced phosphorus removal for municipal wastewater treatment plants. It was observed that anaerobic P release and aerobic P uptake rates for the AS varied from 0.22 to 7.9 mg/g VSS/h and 0.43 to 8.11 mg/g VSS/h, respectively. Moreover, the fraction of Accumulibacter (PAOs: polyphosphate-accumulating organisms) was $4.8 \pm 2.0\%$ of the total biomass, while Competibacter (GAOs: glycogen-accumulating organisms) accounted for $4.8 \pm 6.4\%$. The anaerobic P-release rate was found to be an effective indicator of EBPR. Ji et al. [50] demonstrated the effect of magnetic field on the activity of AS in wastewater treatment. The authors reported the enhanced stimulation of AS acclimatization and organic pollutant biodegradation under the influence of magnetic field which resulted in higher efficiency of wastewater treatment. Beliavski et al. [51] described the denitrification of brines originating from membrane treatment of groundwater in an upflow sludge blanket reactor. The authors came to the conclusion that based on a comparison between two denitrification reactors with and without salt addition and using acetic acid as the electron donor, that the reason for the excellent sludge settling characteristics found in the acetic acid fed reactor is the positive effects of higher salinity on granular sludge formation. AS treatment plant comprises many zones wherein different treatment processes occur. The components and the different processes that occur therein are discussed as below:

7.3.9.1 Anaerobic Zone

This zone is characterized by the absence of both dissolved oxygen and oxidized nitrogen [52]. Herein, the sludge from the clarifier flows jointly with the influent wastewater. It is necessary that oxygen and nitrates must be absent for the proper functioning of this zone. According to the report by Tanaka et al., this zone is responsible for the release of phosphate [53]. Takeda and Shigeki [54] herein described a method for biological denitrification using an anaerobic ammonia oxidation process. In this method, the proportion of the support to be introduced is regulated so that the nitrite-producing nitrification reaction caused by the nitrite-producing nitrifying bacteria yields nitrite in an amount sufficient to inhibit the nitrite-producing nitrification reaction, while keeping the raw water alkaline in the reaction tank. Boopathy [55] developed a laboratory scale SBR method which was successfully operated using shrimp aquaculture wastewater. The wastewater was contaminated with high levels of carbon and nitrogen. Nitrification, denitrification, and the removal of carbon were achieved by operating the reactors sequentially via aerobic and anoxic modes. Moreover, ammonia present in the waste was nitrified within 4 days whereby the denitrification of nitrate was achieved by the anoxic process leading to 100% removal of nitrate within 15 days of reactor operation.

7.3.9.2 Primary Anoxic Zone

The main dentrification reactor in the process is the primary anoxic zone. This component of the plant is supplied with the effluent from the anaerobic zone and mixed liquor recycled from the aerobic zone. Enrichment of denitrifying bacteria is assisted by the presence of nitrate or nitrite and the absence of oxygen. These bacteria are known to reduce nitrate or nitrite to molecular nitrogen. As a consequence, soluble and colloidal biodegradable matter is readily removed in this zone [56]. Zhang et al. [57] reported a three-stage biological denitrification including an organic matter oxidation stage, a nitrification stage, and a denitrifying stage and is characterized by an anaerobic phosphorus release-treating unit additional set before the three-stage biological denitrification and a water inflow mixing unit. The application has the advantages of high denitrification efficiency, improved phosphorus removal capacity, ensured carbon source, low running cost (lowered by 5–10%), reduced occupied area, and saved investment.

7.3.9.3 Primary Aerobic Zone

The primary aerobic zone oxidizes the organic material present in wastewater. In addition, ammonia is converted into nitrate. This zone also provides an environment to take up all the phosphate released in the anaerobic zone [58]. Ammonia must be oxidized to nitrites by nitrifying bacteria such as *Nitrosomonas*, *Nitrosospria*, and *Nitrosolobus* species prior to its removal. Later on, nitrites are oxidized to nitrates by *Nitrobacter*, *Nirtosospira*, and *Nitrococcus* species. Nitrates formed are then removed by denitrifying bacteria in the primary anoxic zone. Enrichment of the AS with bacteria capable of taking orthophosphate determines the phosphate uptake in the process and it is a well known fact *Escherichia coli* aids the enhanced phosphate removal in AS [59]. Jinxia et al. [60] herein described a method for the treatment of aquaculture and eutrophic water bodies using large-sized algae. Large-sized algae clone has the advantages of continuous culture in the whole year, no differentiation, wide temperature endurance range, rapid starting, high breeding, and easy removal and high removal rate for nitrogen and phosphorus. Hongyan et al. [61] reported a method revealing the treatment of wastewater by KM biological catalytic oxidation. They observed that the reported method had the advantages of good ammonia–Nitrogen removal effects, high effluent water quality, and low cost.

7.3.9.4 Secondary Anoxic Zone

The excess nitrate that was not removed in the above zone is converted into nitrogen in this zone. However, the quantity of nitrate removed is very small because of the very slow denitrification rate in this zone. Because of the lower COD, the retention time in the anoxic zone is relatively long [58].

7.3.9.5 Secondary Aerobic Zone and Clarifier

Excess phosphate, which was not removed in the primary aerobic zone, is removed in this zone. In addition residual ammonia undergoes oxidation in this zone. Secondary aerobic zone increases the level of the dissolved oxygen by about 2 to 4 mg/L in the mixed liquor before its entry into the clarifier. This zone requires greater aeration to promote phosphate uptake and to maintain good aerobic conditions. Biomass retains phosphorus as long as aerobic conditions prevail [13]. This zone avoids the development of anaerobic condition in the clarifier and phosphate release before clarification. Treated wastewater which is free of organic matter and dissolved solids is released into the clarifier [37,62]. Tirthakar and Vinaykumar [63] reported the removal of N/P from wastewater during eutrophication. The SBR is a modification of AS process, which has been successfully used to treat municipal and industrial wastewaters. The SBR technology developed in the late 1950s and early 1960s has been used for this purpose. Slusarczyk [64] described a biological wastewater treatment method using a biological reactor. In this method, the processes are conducted in an alternating and joint, aerobic zone, anoxic zone, and aerobic zone arrangement.

7.4 Wastewater Effluent Quality and Regulatory Standards

Biological wastewater treatment is destined to produce effluent that can be used for agricultural or aquacultural activities. However, if the effluent is not suitable for use in agriculture or aquaculture it should be suitable for discharge into inland or coastal waters. Effluent quality standards are determined and set by regulatory agencies empowered by legislation. Unfortunately, such regulations are not as sensible in many countries as they should have been [41]. Wastewater treatment systems should be provided permits from appropriate authorities [65]. For example, discharge from a treatment plant into a stream in the US requires a National Pollutant Discharge Elimination System (NPDES) permit. The NPDES permit is a manifesto that specifies the maximum allowable level of total SS, nutrients, bacteria, BOD, and the minimum level of dissolved oxygen present in the effluent that can be discharged into a stream. NPDES permits are often changed every 5 years since water quality concerns change throughout [65]. Water bodies and soil are approaching toward a serious pollution threat and therefore, it is the need of the hour to control the quality of the effluents mainly with regard to the polluting power (BOD, TOC, SS, and COD) and nutrient contents (phosphate, nitrate, and ammonium). Toxins like solvents, heavy metals, phenols, and chlorinated compounds are also controlled depending on the type of the industry [65].

7.5 Efficiency of Biological Wastewater Treatment

The emergence of certain new challenges has significantly limited the effectiveness of biological wastewater treatment processes over the last few decades [37]. As per the reports by Zhou and Smith, increased knowledge of the consequences of water pollution and the desire of public for better water quality promoted the implementation of strict regulations by expanding the scope of regulated contaminants and lowering their maximum contaminant levels [62]. The rapid population growth, industrial development, and diminishing water resources are also contributing factors [6]. Today's wastewater treatment sector faces several key challenges like old and worn-out collection facilities which require further improvement, replacement, or repair to maintain their useful life; the quantity and character of contaminants are far more complex than those that in the past; population explosion is in demand of creating new plants; increasing urbanization and farm runoff provide additional sources of pollution which are not controlled by conventional wastewater treatment [41]. Depending on the level of treatment, the treatment plants can provide, varying amounts of contaminants from wastewater are removed [7]. Over the years, disinfection techniques like chlorination, UV irradiation, and ozonation have shown varying degrees of success for the removal of pathogens from wastewater [66]. However, inadequacies of these techniques in the removal of certain pathogens from wastewater have been pointed out by the recent literature. Chemical disinfection with chlorine and irradiation by UV radiations have been reported to be ineffective against some viruses and some bacterial spores like *Acanthamoeba*, *Cryptosporidium*, and *Giardia* spp. [67]. For limiting the formation of bromate, ozone applied at low CT (concentration and contact time) was reported to have relatively little effect on the infectivity of the protozoan cysts [9]. Advancement of wastewater treatment technology notwithstanding, treated sewage may still contain some harmful identities irrespective of thoroughness and sophistication of the treatment process, although in smaller quantities than in raw sewage.

7.6 Consequences of Inadequate Wastewater Treatment

Discharge of untreated or inadequately treated wastewater into the environment puts diverse hazardous effects on the overall living systems. Municipal wastewater often results in eutrophication, algal blooms; depleted dissolved oxygen, fish extinction;

sedimentation leading to destruction of aquatic habitats and increased water flow. Moreover, aquatic life suffers from acute and chronic toxicity as well as bioaccumulation and biomagnification of chemicals in the food chain is observed [68,69]. Releasing untreated or inadequately treated municipal wastewater effluents into the environment puts public health at risk from drinking water contaminated with pathogenic bacteria, protozoans (such as *Giardia* and *Cryptosporidium* spp.), and several toxic substances [70]. Moreover, consuming fish and shellfish and engaging in recreational activities in contaminated waters also become mandatory to avoid activity [68,69]. It has been established that carcinogenic and endocrine disrupting substances and some pharmaceuticals can pass through even the most advanced wastewater treatment systems [71]. Naturally occurring hormones are disrupted by endocrine disrupting substances and, therefore, have harsh effects on the growth, reproduction, or development of many wildlife species [72]. According to Goodland and Daly (1996) the natural capital is comprised of intact ecosystems and ecosystem services (structurally and functionally). Wastewater pollution considerably affects the ecosystem; causing depletion of plant biomass, high rate of wetland destruction, drastically affecting aquatic wildlife habitats, and the decrease in freshwater access, the ecosystem services provided by these components will continue to degrade [72]. All these processes deplete the natural capital of the earth and as a result ecosystem loses its capacity to provide the usual vital services. Therefore, implementing a high wastewater effluent standard will maintain the natural capital to ensure that adequate resources are available for natural benefits [73].

7.7 Recent Developments in Biological Wastewater Treatment Methodologies

In order to eradicate the problems (biomass washout, wastewater discharges into freshwater bodies, etc.) that are usually encountered while carrying out wastewater treatment by biological methods; a considerable development has been witnessed in biological wastewater treatment methods recently. Duran et al. [74] during the last decade studied the effects of oxygen on the removal of polymeric resins in upflow anaerobic sludge blanket (UASB) reactor in a biological wastewater treatment process. Two UASB reactors (A and B) were used. Reactor A was used as methanogenic control during the study and both reactors were operated under methanogenic conditions. Reactor A had reduction in COD removal efficiency from $75 \pm 4\%$ to $37 \pm 5\%$, by the increase of polymeric resin compounds (PRC) loading rate from 750 to 1125 mg COD/L. However, there was no styrene elimination in this reactor. In reactor B, the COD removal efficiency was between $73 \pm 5\%$ and $80 \pm 2\%$, even with the addition of O_2 and increase of the PRC loading rate, owing to oxygen being used in the partial oxidation of these compounds. Recently, Zhang et al. [75] reported a wastewater treatment system for the ethanol fermentation industry. The authors demonstrated that the waste was able to be recycled for the next fermentation after being treated with bio-flocculation process.

Moreover, this new process was confirmed to have stable operation over 10 recycles. Furthermore, bio-flocculation was shown to be an effective way to diminish the content of inhibitory compounds (SS, volatile acid, and Fe ions inhibited fermentation) drastically when the waste was recirculated. Robles-Vargas et al. [76] recently described an electrochemical AOPs which resulted in highest toxicity and COD removals (96% for both parameters) in comparison to biological and physicochemical treatments. Adsorption with zeolite and chelex ion-exchange resins and activated carbon lead to a removal of 60–90% of COD and 60–99% toxicity. Ishii et al. [77] recently reported published from a MFC (microbial fuel cell) that was operated for over 300 days using only primary clarifier effluent collected from a municipal wastewater treatment plant as the microbial resource and substrate. The reactor solution was replaced once every 2 weeks with new primary effluent (containing different microbial and chemical compositions with every batch exchange). The authors observed that the turbidity of the primary clarifier effluent solution decreased notably, and 97% of BOD was removed after an 8- to 13-day residence time for each batch cycle. Furthermore, it was very interesting to note that the electrochemical performance and BOD removal rates were very reproducible throughout MFC operation regardless of the sample variability associated with each wastewater exchange. While MFC performance was very reproducible, the phylogenetic analyses of anode-associated electricity-generating biofilms showed that the microbial populations temporally fluctuated and maintained a high biodiversity throughout the yearlong experiment. These results suggest that MFC communities are self-selecting and self-optimizing, thereby able to develop and maintain functional stability regardless of fluctuations in carbon source and regular introduction of microbial competitors. These results contribute significantly toward the practical application of MFC systems for long-term wastewater treatment as well as demonstrating MFC technology as a useful device to enrich for functionally stable microbial populations. Matos et al. [78] in 2012 reported the effect of organic loading rate on the growth of filamentous fungi. High amounts of fungal filaments were observed in the SBR operating at higher organic loading rate, as ascertained by direct microscopic inspection, while at lower organic loading rates, overabundance of fungal filaments was not observed. Sequence retrieved from the isolated fungal filaments presented high similarity (99%) to *Galactomyces geotrichum.* Alturki et al. [79] quite recently reported the potential and several challenges in the development of a novel osmotic membrane bioreactor (OMBR) process for the treatment of municipal wastewater. Following the initial gradual decline, a stable permeate flux value was obtained after approximately 4 days of continuous operation. There was evidence of continuous deterioration of biological activity of the OMBR system, possibly due to the buildup of salinity in the reactor. The removal of 25 of 27 trace organic compounds with molecular weight higher than 266 g/mol was above 80% and was possibly governed by the interplay between physical separation of the FO membrane and biodegradation. In contrast, the removal efficiency values of the other 23 trace organic compounds with molecular weight less than 266 g/mol were very scattered. The removal efficiency of these low molecular weight compounds by OMBR treatment appears to depend mostly on biological degradation. Sui et al. [80] carried out the monthly studies of the occurrence of 12 pharmaceuticals and personal care products (PPCPs) in two wastewater treatment plants in Beijing over the course of 1 year. The authors reported the comparison of the removal of PPCPs by three biological treatment processes viz conventional AS (CAS), biological nutrient removal (BNR), and membrane bioreactor (MBR) during different seasons. Discrepant seasonal variations of PPCPs in the wastewater influent were noted while most PPCPs had lower concentrations in the wastewater effluent, during summer than during winter. The performance of MBR was demonstrated to be more stable for PPCPs than CAS or BNR particularly during winter months. However, pharmaceuticals like diclofenac, trimethoprim, metoprolol, and gemfibrozil could be moderately removed by MBR, while their removal by CAS and BNR was much lower or even negligible. Finally, the authors came to the conclusion that the toxic tank was the most important unit, whereas membrane filtration made a negligible contribution to their elimination. Jagadevan et al. [81] reported a novel hybrid nano zerovalent iron (nZVI)-initiated oxidation—biological degradation approach for remediation of recalcitrant waste metalworking fluids. The authors performed laboratory scale batch studies for the testing of degradation ability of a semi-synthetic metalworking fluid (MWF) wastewater (which was found to be resistant to initial bacterial treatment in specifically established bioreactors) by employing a novel hybrid approach. The main purpose of the approach was to combine the synergistic effects of nZVI-induced oxidation which was followed by biodegradation, specifically for the remediation of recalcitrant components of MWF effluent. The method reported an overall COD reduction of 95.5% demonstrating that nZVI oxidation can be exploited for enhancing the biodegradability of a recalcitrant wastewater in treatment processes. Buth et al. in 2011 [82] described a method for the formation and removal of chlorinated triclosan derivatives in wastewater treatment plants using chlorine and UV disinfection. Wastewater influent, secondary (pre-disinfection) effluent, and final (post-disinfection) effluent samples were collected from two AS wastewater treatment plants. One of the wastewater plants was using chlorine disinfection and the other was using UV disinfection on two occasions. Li et al. [83] described dynamics of functional gene copy number and overall bacterial community during microcystin-LR (MCLR) degradation by a biological treatment facility in a drinking water treatment plant. The authors reported that MCLR effectively biodegraded within 5.5 days in biofilm cultures sampled in winter. Seasonal variations in the MCLR-degradation potential of the biofilm were closely related to the initial MCLR-degrader population in the biofilm as revealed by the quantitative polymerase chain reaction (qPCR). The qPCR assay indicated that MCLR may be a primary substrate for the degraders in the presence of another labile organic carbon associated with the biofilm under the present study conditions. This study can be very helpful in understanding the microbial degradation of MCs and could facilitate the bioremediation of MCs in polluted habitats. Di et al. [84] described a method for tannery wastewater treatment. The authors reported that the method met the present discharge limits when the biologically treated wastewater was recirculated through an adjacent reactor where a specific ozone dose of

120 mg O_3/L (influent) was used. The final effluent of the process looked like tap water which was indicative of the benefits of the process. The proposed method was able to reduce the sludge production by 25–30 times and to save 60% of operating costs in comparison to the conventional methods. Coma et al. [85] reported the effect of cycle changes on biological nutrient removal in a SBR. Their results indicated that the key factors in stabilizing the system for long-term periods with high nutrient removal (88%, 93%, and 99% of carbon, nitrogen, and phosphorus, respectively) are the length of phases and the optimization of influent distribution.

Conclusion

Biological wastewater treatment is one of the frontier areas of research in the present scenario. The sewage after treatment may still contain some harmful entities irrespective of the sophistication and ruggedness of the treatment process. Water which was once regarded as the most abundant natural resource is increasingly becoming one of the scarcest resources now. Recent advances in wastewater treatment methodologies suggest encouraging developments in the treatment systems in near future. Still, a broad spectrum of pathogens happen to occur in wastewater. The numbers and types of such pathogens present are highly dependent on the socioeconomic conditions and customs of the communities creating the wastewater. It is the need of the time to understand the hazardous environmental impacts posed by the untreated or inadequately treated wastewater that enters the nearby ecosystems. In view of the above-summarized facts, it becomes the duty of the environmental scientists all over the world to work for the development of highly efficient and low-cost wastewater treatment methods to ensure safe wastewater treatment for the welfare of the mankind.

References

[1] S. Baccella, G. Cerichelli, M. Chiarini, C. Ercole, E. Fantauzzi, A. Lepidi, L. Toro, F. Veglio, Biological treatment of alkaline industrial waste waters, Process Biochem. 35 (2000) 595–602.

[2] D. Day, How Australian social policy neglects environments, Austr. J. Soil Water Conserv. 9 (1996) 3–9.

[3] V.T. Chow, R. Eliason, R.K. Linsley, Development and trends in wastewater engineering, in: Wastewater Engineering, McGraw-Hill Book Company, New York, 1972, pp. 1–11.

[4] M.R.J. Doorn, S. Towprayoon, S. Maria, M. Vieira, W. Irving, C. Palmer, R. Pipatti, C. Wang, Wastewater treatment and discharge, in: 2006 IPCC Guidelines for National Greenhouse Gas Inventories, vol. 5, (Waste). WMO, UNEP, 2006, pp. 6, 1–6, 28.

[5] M.W. LeChevallier, K. Au, Inactivation (disinfection) processes, in: Water Treatment and Pathogen Control, IWA Publishing, 2004, pp. 41–65.

[6] United States Environmental Protection Agency (U.S. EPA), Primer for Municipal Wastewater Treatment Systems, U.S. EPA, Washington, DC, 2004.

[7] Canada's National Environmental Indicator Series (2003). http://www.ec.gc.ca/soer-ree.

[8] A.R. Upadhyay, V.K. Mishra, S.K. Pandey, B.D. Tripathi, Biofiltration of secondary treated municipal wastewater in tropical city, Ecol. Eng. 30 (2007) 9–15.

[9] W.A.M. Hijnen, E.F. Beerendonk, G.J. Medema, Inactivation of UV radiation for viruses, bacteria, and protozoan (oo) cysts in water: a review, Water Res. 40 (2006) 3–22.

[10] W.A.M. Hijnen, A.J. van der Veer, E.F. Beerendonk, G.J. Medema, Increased resistance of environmental anaerobic spores to inactivation by UV, Water Sci. Technol: Water Suppl. 4 (2) (2004) 54–61.

[11] O. Hoyer, Water disinfection with UV radiation-requirements and realization. in: Proceedings of the European Conference UV Karlsruhe, UV radiation, Effects and technologies, 22–24 September 2003, Karlsruhe, 2004.

[12] H. Schaar, M. Clara, O. Gans, N. Kreuzinger, Micropollutant removal during biological wastewater treatment and a subsequent ozonation step, Environ. Pollut. 158 (5) (2010) 1399–1404.

[13] M. von Sperling, C.A. de Lemos Chernicharo, Biological Wastewater Treatment in Warm Climate Regions, IWA Publishing, London, 2005.

[14] M. Kotsou, A. Kyriacou, K. Lasaridi, G. Pilidis, Integrated aerobic biological treatment and chemical oxidation with Fenton's reagent for the processing of green table olive wastewater, Process Biochem. 39 (2004) 1653–1660.

[15] H.Z. Zhao, P. Cheng, B. Zhao, J.R. Ni, Yellow ginger processing wastewater treatment by a hybrid biological process, Process Biochem. 43 (2008) 1427–1431.

[16] A. Baban, A. Yediler, G. Avaz, S.S. Hostede, Biological and oxidative treatment of cotton textile dye-bath effluents by fixed and fluidized bed reactors, Bioresource Technol. 101 (4) (2010) 1147–1152.

[17] M.M. Ballesteros Martin, J.A. Sanchez Perez, J.L. Garcia Sanchez, J.L. Casas Lopez, S. Malato Rodriguez, Effect of pesticide concentration on the degradation process by combined solar photo-Fenton and biological treatment, Water Res. 43 (15) (2009) 3838–3848.

[18] R. Dewil, J. Baeyens, E. Neyens, Reducing the heavy metal content of sewage sludge by advance sludge treatment methods, Environ. Eng. Sci. 23 (2006) 994–999.

[19] Z. Fu, Y. Zhang, X. Wang, Textiles wastewater treatment using anoxic filter bed and biological wriggle bed-ozone biological aerated filter, Bioresource Technol. 102 (4) (2011) 3748–3753.

[20] L.H. Leal, N. Vieno, H. Temmink, G. Zeeman, C.J. Buisman, Occurrence of xenobiotics in gray water and removal in three biological treatment systems, Environ. Sci. Technol. 44 (17) (2010) 6835–6842.

[21] Y.J. Chan, M.F. Chong, C.L. Law, Biological treatment of anaerobically digested palm oil mill effluent (POME) using a Lab-Scale Sequencing Batch Reactor (SBR), J. Environ. Manage. 91 (8) (2010) 1738–1746.

[22] K. Vijayaraghavan, D. Ahmada, A.Z. Mohd, Aerobic treatment of palm oil mill effluent, J. Environ. Manage. 82 (2007) 24–31.

[23] B.M. Souza, A.C. Cerqueira, G.L. Sant Anna, M. Dezotti, Oil-refinery wastewater treatment aiming reuse by advanced oxidation processes (AOPs) combined with biological activated carbon, Ozone: Sci. Eng. 33 (5) (2011) 403–409.

[24] X.Z. Cao, Y.M. Li, Treatment of linear alkylbenzene sulfonate (LAS) wastewater by internal electrolysis-biological contact oxidation process, Water Sci. Technol. 64 (1) (2011) 147–154.

[25] X. Zhu, Y. Chen, Reduction of N_2O and NO generation in anaerobic-aerobic (low dissolved oxygen) biological wastewater treatment process by using sludge alkaline fermentation liquid, Environ. Sci. Technol. 45 (6) (2011) 2137–2143.

[26] S. Dhadse, S. Satyanarayan, P.R. Chaudhari, S.R. Wate, Vermifilters: a tool for aerobic biological treatment of herbal pharmaceutical wastewater, Water Sci. Technol. 61 (9) (2010) 2375–2380.

[27] M. Oliveira, C. Queda, E. Duarte, Aerobic treatment of winery wastewater with the aim of water reuse, Water Sci. Technol. 60 (5) (2009) 1217–1223.

[28] F. Beline, M.L. Daumer, L. Loyon, A.M. Pourcher, P. Dabert, F. Guiziou, P. Peu, The efficiency of biological aerobic treatment of piggery wastewater to control nitrogen, phosphorus, pathogen and gas emissions, Water Sci. Technol. 57 (12) (2008) 1909–1914.

[29] N. George, L. Andrew, Control of filamentous organisms in food processing wastewater treatment by intermittent aeration and selectors, J. Chem. Technol. Biotechnol. 78 (2003) 420–430.

[30] R.J. Craggs, J.P.S. Sukias, C.C. Tanner, R.J. Davies-Colley, Performance of an Advanced Pond System Treating Dairy Farm Effluent (Incorporating Preliminary Guidelines), NIWA, Hamilton, 2003. M. Elimelech, The global challenge for adequate and safe water, J. Water Supply Res. Technol.-AQUA 55 (2006) 3–10.

[31] Y. Qian, X. Quan, Y. Pei, Z. Yang, Treatment of high ammonia nitrogen polluted river water by biological contact oxidation process, Zhongguo Jishui Paishui 27 (17) (2011) 40–45.

[32] A.A. Stepkin, Y.A. Stepkina, Method and apparatus for aerobic biological treatment of wastewater and complex organomineral fertilizer derived from it. Patent No. RU 2433962, Date: Nov 20, 2011, Russia.

[33] R. Sharman, Process Control II, Water and Waste water Technology, LBCC. Available from: http://www.lbcc.cc.or.us/process2/tricklingfilter/tricklingsum.html, 2004.

[34] G. Tchobanoglous, A.N. Angelakis, Technologies for wastewater treatment appropriate for reuse: potential for applications in Greece, Water Sci. Technol. 33 (1996) 15–24.

[35] R. Al-Saed, Process Performance Optimisation of Al-Bireh Wastewater Treatment Plant, Al-Biereh Municipality, Ramallah, Palestine, 2001.

[36] G. Ji, J. Tong, Y. Tan, Wastewater treatment efficiency of a multi-media biological aerated filter (MBAF) containing clinoptilolite and bioceramsite in a brick-wall embedded design, Bioresource Technol. 102 (2) (2011) 550–557.

[37] P. Smeets, L. Rietveld, W. Hijnen, G. Medema, T. Stenström, Efficacy of water treatment processes, in: MICRORISK (Microbiological Risk Assessment: A Scientific Basis for Managing Drinking Water Safety from Source to Tap), April 2006. http://217.77.141.80/clueadeau/microrisk/uploads/microrisk efficacy of water treatment processes.

[38] A. Tawfit, B. Klapwijk, F. El-Gohary, G. Lettinga, Treatment of anaerobically pre-treated domestic sewage by rotating biological contractor, Water Res. 36 (2002) 147–155.

[39] F. Kargi, Empirical models for biological treatment of saline wastewater in rotating biodisc contactor, Process Biochem. 38 (2002) 399–403.

[40] T. Arnaud, Treatment of winery wastewater with an anaerobic rotating biological contactor, Water Sci. Technol. 60 (2) (2009) 371–379.

[41] D.D. Mara, Domestic Wastewater Treatment in Developing Countries, Earthscan Publications, London, 2004.

[42] F. Xiang, Z. Jiti, Method for completely recycling wastewater treatment sludge. Patent No. 102206025, October 5, 2011.

[43] A. Kawae, Y. Tani, H. Nakamura, A. Tanaka, H. Miyabayashi, Method for treating paper manufacturing wastewater by activated-sludge process using phosphorus component as microorganism nutrient, Jpn. Kokai Tokkyo Koho (2011). JP 2011230044 A 20111117.

[44] S.J. Park, T.I. Yoon, J.H. Bae, H.J. Seo, H.J. Park, Biological treatment of wastewater containing dimethyl sulphoxide from the semi-conductor industry, Process Biochem. 36 (2001) 579–589.

[45] Fikret Kargi, Serpil Ozmıhc, Performance of azotobacter supplemented activated sludge in biological treatment of nitrogen deficient wastewater, Process Biochem. 38 (2002) 57–64.

[46] J. Bohdziewicz, E. Sroka, Integrated system of activated sludge–reverse osmosis in the treatment of the wastewater from the meat industry, Process Biochem. 40 (2005) 1517–1523.

[47] H. Li, Z. Zhang, H. Wang, X. Xu, Y. Lin, J. Zhang, Z. Jun, Biological phosphorus removal performance and relevant microorganism characteristics of activated sludge in municipal wastewater treatment plants, China, Water Environ. Res. 83 (7) (2011) 594–602.

[48] D. Zhang, W. Li, S. Zhang, M. Liu, X. Zhao, X. Zhang, Bacterial community and function of biological activated carbon filter in drinking water treatment, Biomed. Environ. Sci. 24 (2) (2011) 122–131.

[49] H.Z. Li, J. Zhu, L. Weiping, X. Xin, Improvement strategy on enhanced biological phosphorus removal for municipal wastewater treatment plants: full-scale operating parameters, sludge activities, and microbial features, Bioresource Technol. 102 (7) (2011) 4646–4653.

[50] Y. Ji, Y. Wang, J. Sun, T. Yan, J. Li, T. Zhao, X. Yin, C. Sun, Enhancement of biological treatment of wastewater by magnetic field, Bioresource Technol. 101 (22) (2010) 8535–8540.

[51] M. Beliavski, I. Meerovich, S. Tarre, M. Green, Biological denitrification of brines from membrane treatment processes using an upflow sludge blanket (USB) reactor, Water Sci. Technol. 61 (4) (2010) 911–917.

[52] D.H. Eikelboom, A. Draaijer. Activated Sludge Information System (ASIS) (1999). http://www.asissludge.com.

[53] Y. Tanakaa, A. Yatagaib, H. Masujimac, M. Wakia, H. Yokoyama, Autotrophic denitrification and chemical phosphate removal of agroindustrial wastewater by filtration with granular medium, Bioresource Technol. 98 (4) (2007) 787–791.

[54] S. Takeda, Method for biological denitrification using anaerobic ammonia oxidation reaction, PCT Int. Appl. (2011). WO 2011148949 A1 20111201.

[55] R. Boopathy, Biological treatment of shrimp production wastewater, J. Ind. Microbiol. Biotechnol. 36 (7) (2009) 989–992.

[56] Metcalf, Eddy, Wastewater Engineering: Treatment and Reuse, in: G. Tchobanoglous, F.L. Burton, H.D. Stensel (Eds.), Tata McGraw-Hill Publishing Company Limited, fourth ed., New Delhi, India, 2003.

[57] C. Zhang, W. Wang, J. Zhang, H. Liu, B. Liu, W. Ye, Three-stage wastewater biological denitrification treating method and apparatus comprising anaerobic phosphorus release treating unit, Faming Zhuanli Shenqing (2012), CN 102329058 A.

[58] H. Torpak, (2006). Activated sludge http://web.deu.edu.tr/atiksu/toprak/ani406.html.

[59] Sci-Tech. Encyclopaedia (2007). http://www.answers.com/topic/sewagetreatment.

[60] J. Wang, R. Dong, B. Zhou, Method for treatment of aquaculture and eutrophic water bodies using large-sized algae, Faming Zhuanli Shenqing (2011), CN 102161530 A 2011082.

[61] H. Wang, J. Wang, X. Chen, Method for treating wastewater by KM biological catalytic oxidation, Faming Zhuanli Shenqing (2011), CN 102285742 A 20111221.

[62] H. Zhou, D.W. Smith, Advanced technologies in water and wastewater treatment, J. Environ. Eng. Sci. 1 (2002) 247–264.

[63] S.N. Tirthakar, C.H. Vinaykumar, Nutriments removal by sequencing batch reactors: an integrated approach in sewage treatment. Part-I, J. Inst. Public Health Eng. 2011 (1) (2012) 36–41.

[64] J.R. Slusarczyk, Method of biological waste-water treatment, Can. Pat. Appl. (2011), CA 2744602 A1 20111224.

[65] World Health Organization, Guidelines for the Safe Use of Wastewater, Excreta and Greywater: Wastewater Use in Agriculture, World Health Organization, Geneva, 2006.

[66] Environmental Protection Agency (EPA), Victoria, Guidelines for Environmental Management: Disinfection of Treated Wastewater. Publication 730, Victoria, Australia, 2002.

[67] J.A. Tree, M.R. Adams, D.N. Lees, Chlorination of indicator bacteria and viruses in primary sewage effluent, Appl. Environ. Microbiol. 69 (2003) 2038–2043.

[68] J. Kapitain, Ontario's Sewage Treatment Plants and Their Effect on the Environment, Environment Probe Publications, www.environmentprobe.org/enviroprobe, 1995.

[69] D.F. Boesch, R.B. Brinsfield, R.E. Magnien, Chesapeake Bay eutrophication: scientific understanding, ecosystem restoration, and challenges for agriculture, J. Environ. Qual. 30 (2002) 303–320.

[70] D. Paillard, V. Dubois, R. Thiebaut, F. Nathier, E. Hoogland, P. Caumette, C. Quentine, Occurrence of *Listeria* spp. in effluents of French urban wastewater treatment plants, Appl. Environ. Microbiol. 71 (11) (2005) 7562–7566.

[71] T. Heberer, Occurrence, fate, and removal of pharmaceutical residues in the aquatic environment: a review of recent research data, Toxicol. Lett. 131 (2002) 5–17.

[72] T. Furuichi, K. Kannan, J.P. Giesy, S. Masunaga, Contribution of known endocrine disrupting substances to the estogenic activity in Tama River water samples from Japan using instrumental analysis and *in vitro* reporter gene assay, Water Res. 38 (2004) 4491–4501.

[73] D.M. Smith, P.C. Roebeling, A.J. Webster, M. Kragt, I. Bohnet, A. Zull, Assessment of the Socio-Economic Impacts of Management Options for Improving Water Quality in Douglas Shire, CSIRO Sustainable Ecosystems, Alice Springs, Atherton, Brisbane, Canberra, Darwin, Perth, Toowoomba, Townsville, 2005.

[74] U. Duran, O. Monroy, J. Gomez, F. Ramirez, Biological wastewater treatment for removal of polymeric resins in UASB reactor: influence of oxygen, Water Sci. Technol. 57 (7) (2008) 1047–1052.

[75] W. Zhang, R. Xiong, G. Wei, Biological flocculation treatment on distillery wastewater and recirculation of wastewater, J. Hazard. Mater. 172 (2–3) (2009 Dec 30) 1252–1257.

[76] D. Robles-Vargas, S. Margarita Montoya-Castillo, F.J. Avelargonzalez, J. Jauregui-Rincon, F.J. Rodríguez-Valadez, R. Rico-Martínez, Assessment of the quality and toxicity of the discharges of a wastewater treatment plant and alternatives to improve its operation, J. Environ. Sci. Health A: Tox. Hazard. Subst. Environ. Eng. 47 (2012) 589–597.

[77] S. Ishii, S. Suzuki, T.M. Norden-Krichmar, K.H. Nealson, Y. Sekiguchi, et al., Functionally stable and phylogenetically diverse microbial enrichments from microbial fuel cells during wastewater treatment, PLoS One 7 (2) (2012) e30495.

[78] M. Matos, M.A. Pereira, A. Nicolau, A.L. Rodrigues, A.N.G. Brito, R. Nogueira, Influence of the organic loading rate on the growth of *Galactomyces geotrichum* in activated sludge, J. Environ. Sci. Health A 47 (2012) 565–569.

[79] A. Alturki, J. McDonald, S.J. Khan, F.I. Hai, W.E. Price, L.D. Nghiem, Performance of a novel osmotic membrane bioreactor (OMBR) system: flux stability and removal of trace organics. Bioresource Technol., 2012 (In Press).

[80] Q. Sui, J. Huang, S. Deng, W. Chen, G. Yu, Seasonal variation in the occurrence and removal of pharmaceuticals and personal care products in different biological wastewater treatment processes, Environ. Sci. Technol. 45 (2011) 3341–3348.

[81] S. Jagadevan, M. Jayamurthy, P. Dobson, I.P. Thompson, A novel hybrid nano zerovalent iron initiated oxidation – biological degradation approach for remediation of recalcitrant waste metal-working fluids, Water Res. 46 (7) (2012) 2395–2404.

[82] J.M. Buth, M.R. Ross, K. McNeill, W.A. Arnold, Removal and formation of chlorinated triclosan derivatives in wastewater treatment plants using chlorine and UV disinfection, Chemosphere 84 (9) (2011) 1238–1243.

[83] J. Li, K. Shimizu, M. Utsumi, T. Nakamoto, M.K. Sakharkar, Z. Zhang, N. Sugiura, Dynamics of the functional gene copy number and overall bacterial community during microcystin-LR degradation by a biological treatment facility in a drinking water treatment plant, J. Biosci. Bioeng. 111 (6) (2011) 695–701.

[84] C. Di Iaconi, G. Del Moro, M. De Sanctis, S. Rossetti, A chemically enhanced biological process for lowering operative costs and solid residues of industrial recalcitrant wastewater treatment, Water Res. 44 (12) (2010) 3635–3644.

[85] M. Coma, S. Puig, H. Monclus, M.D. Balaguer, J. Colprim, Effect of cycle changes on simultaneous biological nutrient removal in a sequencing batch reactor (SBR), Environ. Technol. 31 (3) (2010) 285–294.

8

Perspectives of Water Treatment, Remediation and Recycling

CHAPTER OUTLINE

8.1 Introduction

Water is the most important and essential component of this universe and plays a vital role in the proper functioning of the Earth's ecosystems. In spite of this, safe drinking water is not available in many parts of the world. The quality of water resources is deteriorating exponentially due to their contamination. Both point and nonpoint sources are polluting our water resources due to geometrical growth of population, modern industrialization, civilization, domestic and agricultural activities and other geological, environmental and global changes. Nowadays, water pollution is a serious issue as it affects our lives and is expected to get worse over coming decades [1–3].

Because of the utmost importance of water the need of its quality improvement and preservation is growing continuously. By 2020, the global population is suppose to reach up to 7.9 billion [4] and because of this the world may be under great fresh water scarcity. Our water resources are limited and under such circumstances water recycling and treatment technologies are the only alternatives to fulfill water demand in coming few decades. Therefore, there is a great need for the development of the suitable, inexpensive and rapid wastewater treatment, remediation and recycling or conservation methods in the present century [5,6]. The benefits of water reuse in industries are supplemental and reliable water source to enhance and replace existing water supplies. Water recycling is increasingly practiced throughout the world for various purposes; particularly in arid environments. Therefore, attempts have been to assess the future of wastewater treatment, remediation and recycling methodologies. Efforts have also been carried out to

Environmental Water, http://dx.doi.org/10.1016/B978-0-444-59399-3.00008-8

present a systematic approach for water treatment and recycling methods, which involve the understanding, evaluating and selecting parameters. A comparison of the technologies has been presented by discussing their performance, sludge production, life period and operations.

8.2 Water Recycling

Water recycling is not a new concept but it has emerged in ancient time in China and Egypt due to irrigation requirements. But in fact water is recycled by nature through hydrological cycle since the existence of the earth. First attempts for water reuse was in USA in 1940s when chlorinated domestic wastewater was used in the steel industry [7]. But water recycling started in last quarter of the twentieth century; in USA and European countries. From 1930 to 1970, the amount of reused water in Sweden increased by six times. In Japan (1951), a program for recycling of water of the Mikawashima Wastewater Treatment Plant in Tokyo was started. The recycled water was used for a paper mill. It was observed that the water quality of recycled water was better than the quality of any other available water source. The growing economy of Japan in this period resulted in a competition between industry and agriculture for available water, which made water recycling more important. Today, about eight percent water in industries in Japan is recycled. Nowadays, water recycling and reuse are being carried out throughout the whole world including America, Europe, Africa, Asia and Australia continents. A survey on municipal wastewater reclamation depicted the order of recycling as Japan > USA > Europe > Australia [8]. Japan and Singapore are the most prominent water recyclers in Asia with 150 GL water recycled annually. The success of water reuse depends on many factors such as local communities, companies and other rules and regulations. According to Bixio et al. [8], the applications of recycled water are agricultural irrigation, urban, recreational, environmental uses (recharge, process water for industry, direct and indirect potable water production) and combinations of all these.

8.3 Water Recycling Considerations

Basically, water recycling is the need of the urbanization and the types of recycled water and management influence the water cycle balance. The performance of the water cycle system is based on environmental management and issues, water cycle infrastructure and quantity and quality needs of the different receiving environments. The considerable environmental issues are salinity, soil contamination, public health, water quantity and quality constraints, downstream water infrastructure issues and recycling rainwater/ storm waters. Besides, social and economical implications are also responsible for water recycling and reuse. BASIX is a web-based sustainability tool that is currently used to reduce water consumption for residential developments up to 50%. The introduction of BASIX is an important step to promote the use of recycled water for new residential developments. But, nowadays, scientist are also considering many other issues like total water cycle performance, regulatory guidelines and broadening the criteria for sustainability to account for social and environmental objectives.

8.4 Water Recycling and Health

Recycled waters are being used for irrigation, industrial and domestic supplies. Recycled water requires proper treatment prior to supply. The treatment should be as per the requirement of the supply. Treatment should be capable to remove contaminants that may be harmful to health and wealth. The removal of microbes and toxic metal ions is necessary as these cause serious side effects and, sometimes, are carcinogenic too. The level of treatment requires as per the potentiality of risk. It has been observed that pathogens can be removed effectively by using optimum dose of chlorination. Besides, for some specific purposes, UV radiation and reverse osmosis methods are more effective for providing safe recycled water. For the removal of toxic metal ions, several methods such as adsorption, electrochemical and membrane filtration are available. The application of a particular method depends on the type of recycled water and its use. The detail about these techniques have already discussed in various chapters of this book. The benefits of water recycling include the followings.

1. If the treatment of wastewater is done in such a way that the treated water contains various nutrients essential for plant growth, so it can reduce the use of fertilizers and manure.
2. The treated wastewater can be used in various urban and semi-urban activities like providing water for parks, wildlife refuges, fire fighting, and toilet flushing etc.
3. The untreated discharge of wastewater into various water bodies destroys the quality of the water and affects the flora and fauna of the water body.
4. The use of treated wastewater decreases the load of freshwater for agriculture.
5. Usage of treated wastewater would reduce the discharge of untreated wastewater into the environment so reducing water pollution and the contamination of drinking water bodies.

8.5 Water Recycling Techniques

Generally, all techniques available have been used for recycling wastewaters and the most important include adsorption, electrochemical and biological methods, membrane filtration and disinfection. The selection of the suitable technique depends on the quality of wastewater being treated. For example, membrane technologies and adsorption can be used for any type of wastewater while electrochemical and biological methods are selective. These techniques are being used worldwide for wastewater recycling, i.e. USA, Australia, Japan, Singapore and other European countries. Besides, ultraviolet and hydrogen peroxide disinfection (advanced disinfection) is supposed as the best approach

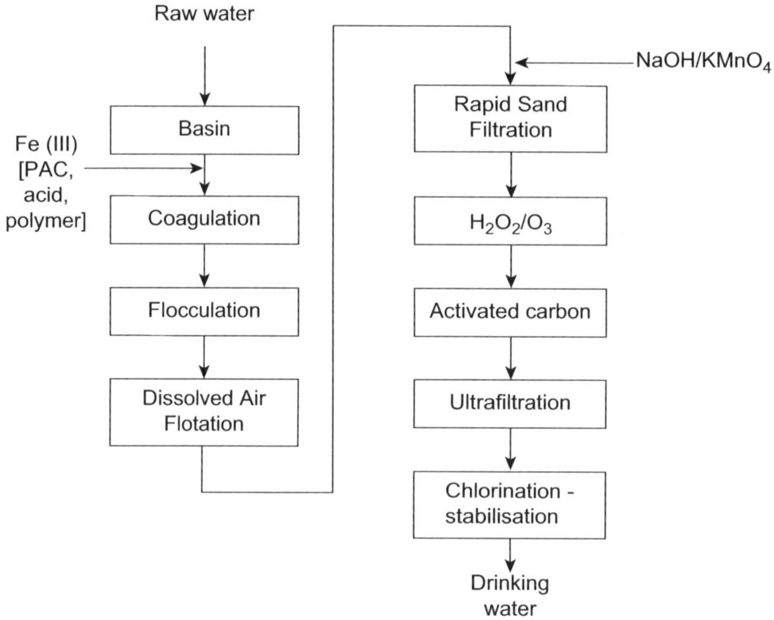

FIGURE 8-1 The most common water recycling scheme.

to make water health risk free. Water recycling is carried out in a systematic and scientific way. The most common water recycling scheme is shown in Figure 8-1.

8.6 Future Perspectives of Water Recycling

As discussed above, the population of the world is increasing continuously and the expected populations of the world in the near future are shown in Figure 8-2. In 2050, the

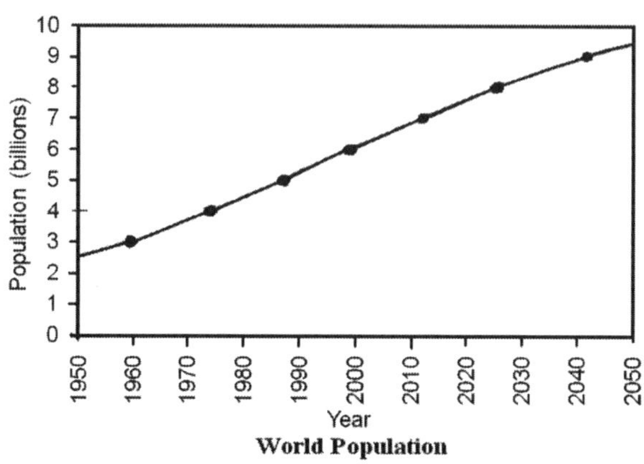

FIGURE 8-2 Expected world population.

expected population of the world may reach up to 9.5 billion. To fulfill the requirements of this population, there will be the growth in agriculture and industrial sectors. All these factors, i.e. increase in population, industrial and agriculture growth will results into an increase in fresh water supply. Of course, the quantity of fresh water on this planet is fixed and the increasing demand in the future may be filled by wastewater recycling and reuse. At present many countries are using water recycling but in the future whole world will depend on this technology. Today, water recycling methodologies are not cost effective and cannot be undertaken by developing and underdeveloped countries. Therefore, there is a great need to develop more economic methods of water recycling so that the underdeveloped countries can use water recycling in the future. Briefly, water recycling is becoming an important part in our lives and will be the need of future.

Conclusion

Nowadays, the treatment of wastewater and recycling are well accepted throughout the world due to fresh water scarcity. The technologies for producing clean water from wastewater are also well developed and available. Besides, the future of recycling is also dependent on awareness and education in public. Of course, water recycling is being done well in developed countries but has not achieved good reputation in developing countries. It is due to the lack of infrastructures required and awareness in people.

References

[1] L.B. Franklin, Wastewater Engineering: Treatment, Disposal and Reuse, McGraw Hill Inc., New York, 1991.

[2] R.L. Droste, Theory and Practice of Water and Wastewater Treatment, John Wiley & Sons, New York, 1997.

[3] D.Z. John, Hand book of Drinking Water Quality: Standards and Controls, Van Nostrand Reinhold, New York, 1990.

[4] T. Dyson, Population and Food: Global Trends and Future Prospects, Routledge, London, 1996.

[5] R.L. Droste, Theory and Practice of Water and Wastewater Treatment, John Wiley & Sons, Inc., New York, 1997.

[6] L.B. Franklin, Wastewater Engineering: Treatment. Disposal and Reuse, McGraw Hill, Inc., New York, 1991.

[7] A.D. Levine, T. Asano, Water reclamation, recycling and reuse in industry, in: P. Lens, L. Hulshoff Pol, P. Wilderer, T. Asano (Eds.), Water Recycling and Resource Recovery in Industry, IWA Publishing, London, UK, 2002.

[8] D. Bixio, B. De heyder, H. Cikurel, M. Muston, V. Miska, D. Joksimovic, A.I. Schäfer, A. Ravazzini, A. Aharoni, D. Savic, C. Thoeye, Municipal wastewater reclamation: where do we stand? An overview of treatment technology and management practice, Water Sci. Technol.: Water Supply 5 (1) (2005) 77–85.

Index

Note: Page numbers followed by "f" indicate figures and "t" indicate tables.

Edwards Brothers Malloy
Ann Arbor MI. USA
July 15, 2013